HELICAL SAIL WINDMILL CONSIDERATIONS

APPLICATIONS:

WATER LIFTING AND PUMPING
IRRIGATION
OPERATION OF HAND-POWERED AGRICULTURAL PROCESSING MACHINERY SUCH AS WINNOWER OR THRESHER

SKILLS/LABOR/TIME:

Construction time and labor resources required to complete this project will vary depending on several factors. The most important consideration is the availability of people interested in working on this project. The project may in many circumstances be a secondary or after work project. This will of course increase the length of time needed to complete the project. The construction times given here are at best an estimation based on limited field experience.

Skill divisions are given because some aspects of the project require someone with experience in metalworking and/or welding. Make sure adequate facilities are available before construction begins.

SKILLED LABOR -- 6 hours
UNSKILLED LABOR -- 30 hours
WELDING -- 1 hour

SPECIAL CONSIDERATIONS:

Advantages:

- Low cost
- Easy to build
- Simple operation

Disadvantages:

- Requires attention when operating
- Designed to be used where wind blows from one or two directions on a regular basis
- Requires the use of metal bearings which may not be available locally
- Canvas sails need replacement periodically

COST ESTIMATE:* $95.00 (US) including labor and pump mechanism.

*Cost estimates serve only as a guide and will vary from country to country.

I. INTRODUCTION

This manual introduces a windmill suitable for construction in countries where the wind is strong and its directions predictable. The cost of production in the country of origin, the Philippines, is approximately $45 US, thus making this wind machine far more feasible financially, for many, than a model which may cost, on an average, $1500.

Simple and Easy to Build!

This advertisement for the windmill appeared in the Philippines Free Press, September 13, 1968.

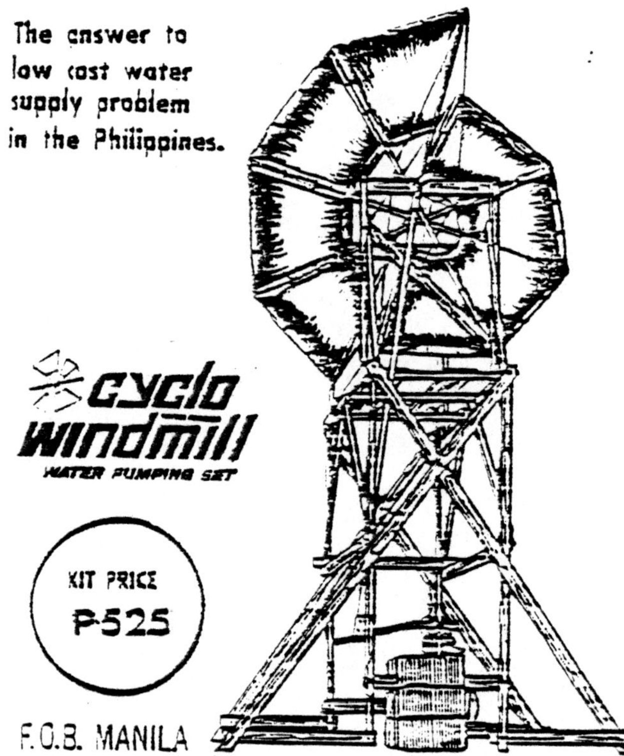

CYCLO WINDMILL water pumping set gets you water anytime, anywhere. Its simple design of screw type helical vanes makes the sails rotate with wind from any direction. You are assured of continuous water flow--from 8 up to 10 gallons per minute.

Most ideal for domestic (household) water needs, irrigation needs for small areas or plots, or for dairy, piggery or poultry needs!

Simple and easy to build, the CYCLO WINDMILL pumping set is sold as a kit complete with component parts, bolts, cables, clips, and shallow well pump with foot-valve, except water piping and wood parts for the tower. A deep well pump cylinder is available at an extra cost.

The windmill was originally fabricated by a Peace Corps Volunteer in 1963. At that time it was know as the Lewis Sawali Windmill. Although working models were built, there were many structural weaknesses. In 1968, another Peace Corps Volunteer, with financial help from the mayor of Santa Barbara, Iloilo, Philippines, constructed the windmill following the same design but substituting stronger parts and prefabricated sections. The windmill has been known since that time as the Santa Barbara Windmill. Plans for the windmill have been distributed around the world.

The windmill, as presented here, has helical sails, or fans, made of heavy canvas with a diameter of 305cm. The tower is made of wood and consists of a double "A-frame" about 488cm high on a concrete base about 305cm square. The axle is steel; the spokes are iron angle bars. The windmill is most suitable in areas where the wind blows along a single directional axis, such as in the Philippines where the wind comes from either the northeast or the southwest.

II. PRE-CONSTRUCTION CONSIDERATIONS

Application

This is a water-pumping windmill which can be placed over a well to provide continuous pumping, or it can be used to pump water from a river to a given site. It can provide a continuous flow of about 8-11 gallons per minute. This is usually an adequate amount of water for domestic and small-farm use.

While it is relatively inexpensive and easy to construct, the windmill does have several possible disadvantages which should be considered very carefully. The first is that provision must be made so that the pump can be used to pump only what is needed and then be disconnected; otherwise, it would continue and could even pump the well dry.

The second potential disadvantage, at least in some geographical areas, is that the mill has no braking system or feathering mechanism to turn it out of the wind. In places such as the Philippines where the weather is most often very steady, the problem is not too great. But if heavy winds or storms are predicted, it will be necessary to remove the sails from the ribs so that the mechanism can turn freely in the wind.

Well before beginning construction, it is important to know how much water must be pumped and for what purposes. If more water (than 11 gallons per minute) is needed, a different windmill probably should be constructed. In addition, the purpose and the quantity needed affect the choice of the pump and the cost of the entire effort.

If digging a well is necessary as part of the effort, there are excellent plans available to assist. Please note that the type of well can and does influence the type of windmill needed; the combination of well and windmill influences the choice of pump. And so on. All of these factors should be considered carefully before construction begins.

Pumps

One good pump for use with this windmill is a self-priming double-action piston pump. The pump handle is attached to the windmill by an eccentric cam rod.

Site Selection

As mentioned earlier, the windmill can be placed over the water source, or it can power a pump connected to a river by a pipe. The area surrounding the windmill should be free of all obstacles that might interfere with the wind--the clear area should extend at least 30 meters. If obstructions are impossible to avoid, the windmill tower can be made higher. For each 91.5cm of additional height required, add 91.5cm to the connecting rod and 99cm to each leg. The foundation should be spaced 30.5cm further apart. No further adjustment is required.

If the windmill is to face a northeast-southwest direction, then the pump assembly must be set up so that the pump handle operates on a north<u>west</u>-south<u>east</u> axis, perpendicular to the wind direction. If possible, the pump should be decided upon before construction begins. On the other hand, the concrete foundation is easier to construct after the windmill is completed.

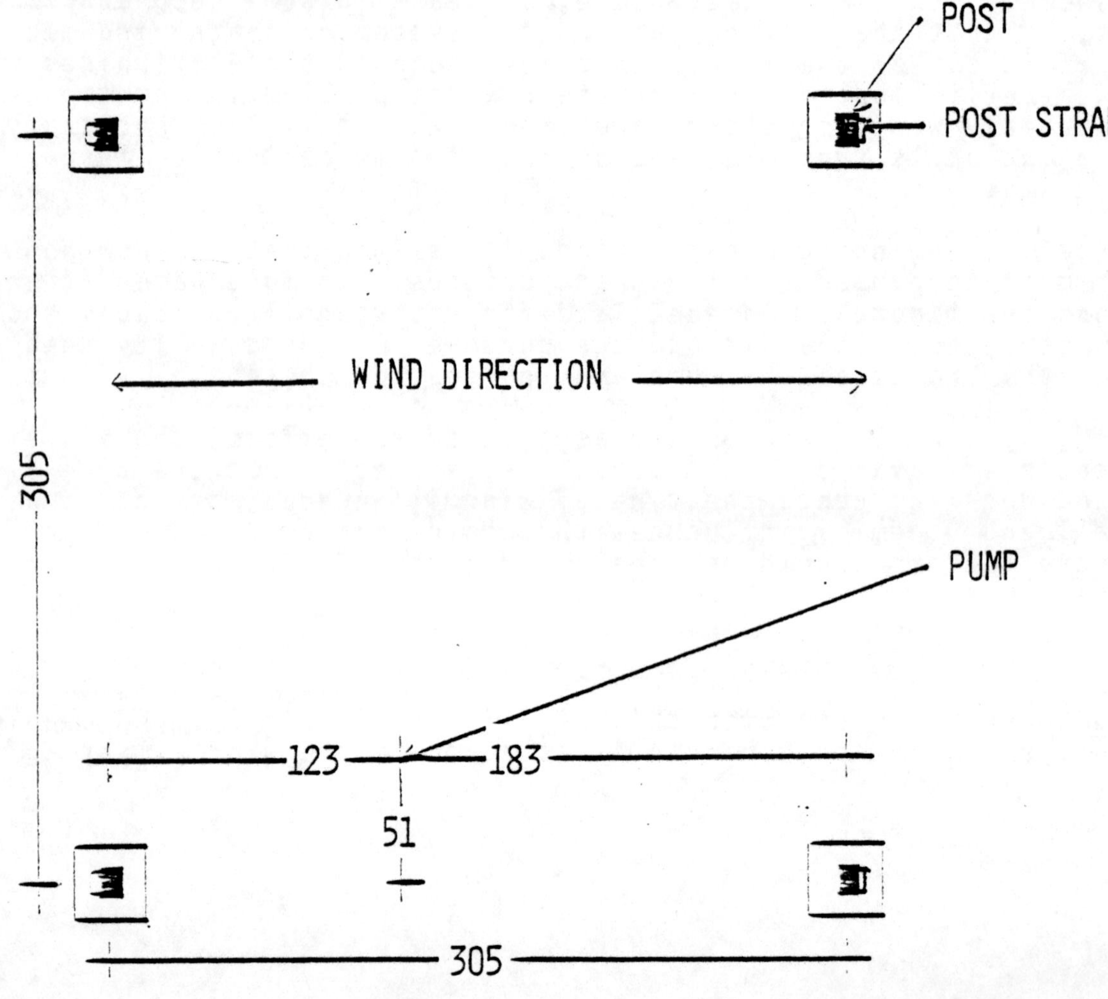

FOUNDATION PLAN

ALL DIMENTIONS IN CENTIMETERS

TOOLS AND MATERIALS

CEMENT FOOTINGS--Grade "B" concrete 1:2-1/2:5

- -- 2 bags cement
- -- 8.8 cubic feet (or 1/4 cubic meter) sand
- -- 17.7 cubic feet (or 1/2 cubic meter) gravel
- -- 4 pieces 6.5mm x 5cm x 61cm post straps

WOOD--Grade 3 lumber except where specified:

- -- 4 pieces, 7.5cm x 10cm x 488cm (for posts)
- -- 2 pieces, 2.5cm x 10cm x 427cm (for braces)
- -- 6 pieces, 2.5cm x 10cm x 366cm (for braces--2 pieces optional)
- -- 1 piece, 5cm x 5cm x 488cm grade 2 lumber (for connecting rod)

BOLTS--Complete with nut and washer unless otherwise specified:

- -- 20 pieces, cap screws 3/8" x 1" (9.5mm x 2.5mm) w/lock washers & double nuts (spoke)
- -- 16 pieces, carriage bolts 3/8" x 4" (9.5mm x 10cm) (Frame A)
- -- 16 pieces, carriage bolts 3/8" x 5" (9.5mm x 12.5cm) (Frame B)
- -- 8 pieces, carriage bolts 3/8" x 2" (9.5mm x 5cm) (Frame C)
- -- 8 pieces, machine bolts 1/2" x 10" (1.25cm x 25.4cm) plus 8 extra washers (Frame D)
- -- 3 pieces, machine bolts 3/8" x 3" (9.5mm x 7.5cm) (E) (locking bolts)
- -- 8 pieces, carriage bolts 3/8" x 4" (9.5mm x 10cm) (F) (footings)

NOTE: If optional bar brace is used, add 4 more 3/8" (9.5mm) diameter x 4" (10cm) carriage bolts and 3/8" (9.5mm) diameter x 5" (12.5cm) carriage bolts each or use nails to secure.

NOTE: In many parts of the world, nuts and bolts may be available only in English measurements.

HARDWARE

- -- 2 pieces, 2.5cm self-aligning ball-bearings with revolvable case
- -- 2 pieces, bearing bolts 10cm long
- -- 1 piece, 2.5cm cold-rolled steel rod, 8' (244cm) long (axle)
- -- 1 piece, 1/2" x 10' (1.25cm x 254cm) ordinary round bar (angle braces)
- -- 3 pieces, 1/4" x 20' (6.5mm x 508cm) ordinary round bar (ribs)
- -- 2.2 lbs. (or 1kg) #16 wire
- -- 3 pieces, 1/8" x 1" x 1" x 20' (3mm x 2.5cm x 2.5cm x 508cm) angle iron bar
- -- 10 pieces, "U" bolts (or jeep spring clips) 1-1/2" (3.8cm) inside diameter
- -- 1 piece 3/8" x 2" x 36" (9.5mm x 5cm x 90.5cm) flat bar
- -- 11.5 meters 91.5cm width heavy weight canvas or tent cloth
- -- 88 pieces canvas grommets (eyelets), #1 or #2
- -- 28.5 meters hemp rope or jute, 6.5mm
- -- 2 gallons coal tar - creosote
- -- 1 pint lead paint
- -- 1.5 liters ordinary paint or canvas protector
- -- Nails

FOR CONNECTING ROD

-- 1 piece, machine bolt 5/8" (1.6cm) diameter x 3" (7.5cm) w/3 nuts and 1 lock washer
-- 1 piece, ball bearing assembly, Honda part #6003 (see text for substitution)
-- 1 piece, 1" (2.5cm) <u>inside</u> diameter pipe, 10cm long
-- 6 pieces, 3/8" x 3" (9.5mm x 7.5cm) machine bolts w/nut (coarse thread)

PUMP

-- 1 self-priming double-action piston pump
-- Materials for plumbing and reservoir as required.

TOOLS

Hand drill

Wood and metal bits

Wrenches

Pliers

Punch (for eyelets)

Hammer

Level

Saw

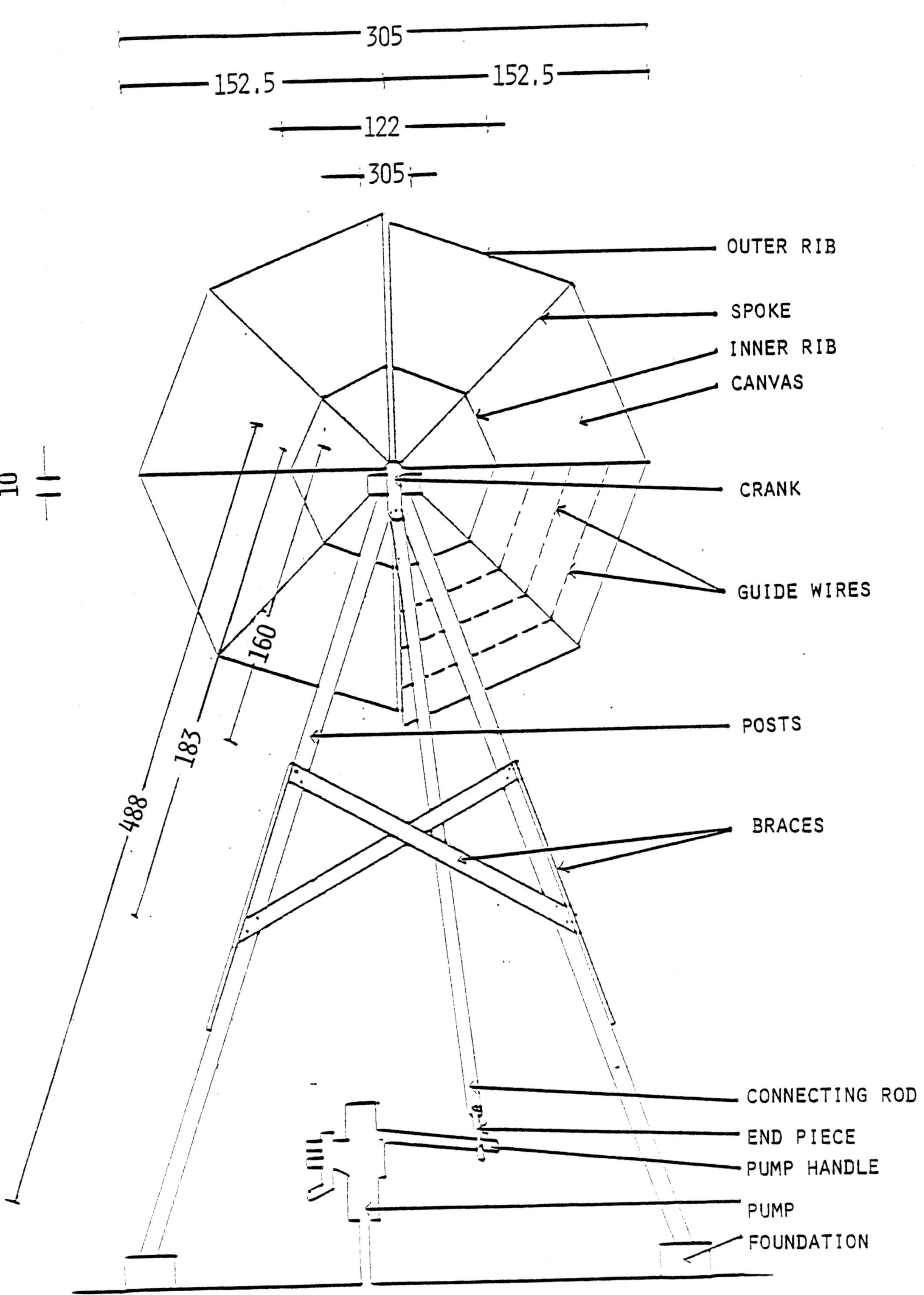

ALL DIMENSIONS IN CENTIMETERS

FRONT VIEW

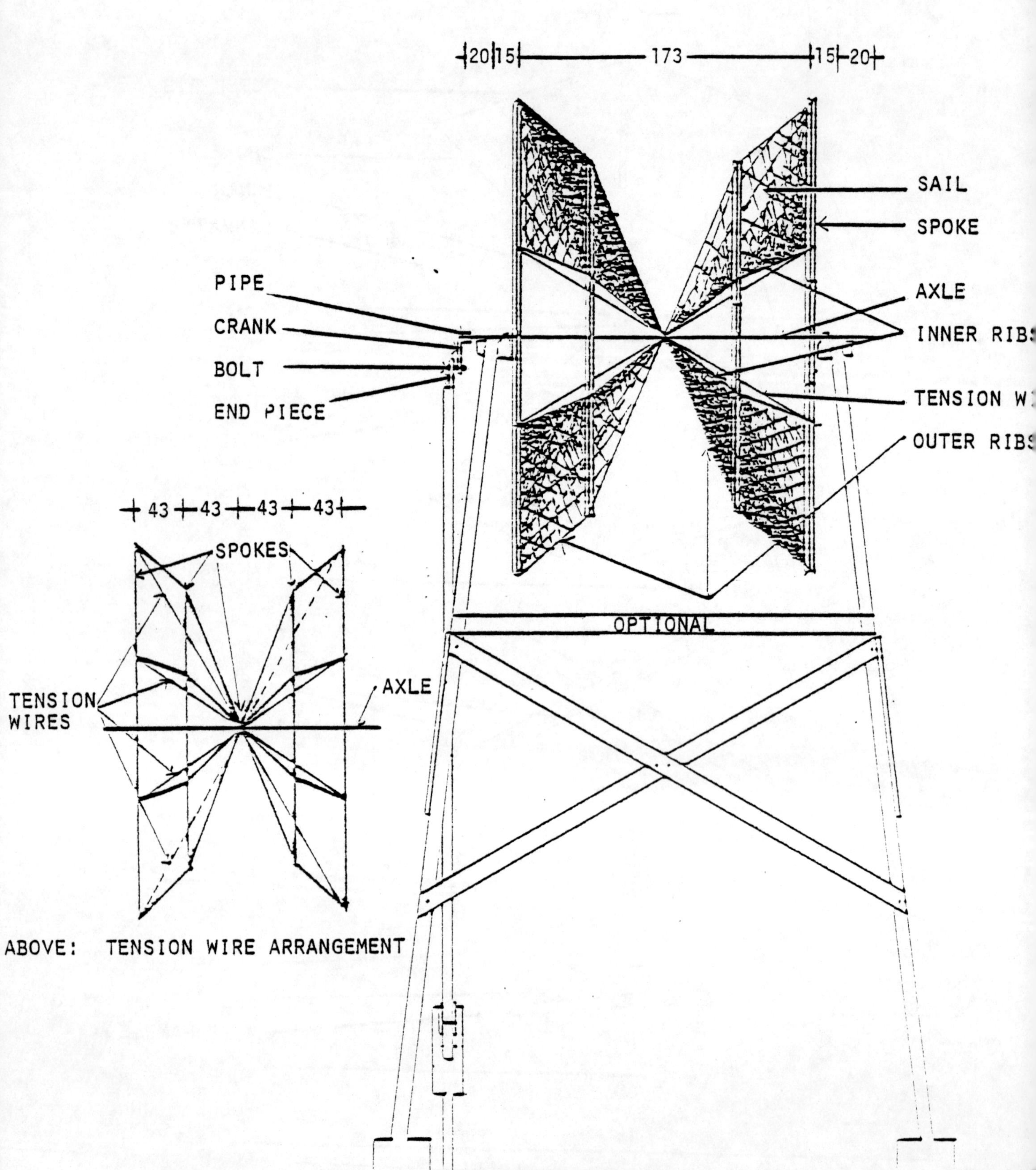

ABOVE: TENSION WIRE ARRANGEMENT

ALL DIMENSIONS IN CENTIMETERS

III. CONSTRUCTION

Build The Frame

-- Cut 30.5cm off the end of each of the four 7.5cm x 10cm x 488cm wood posts. Save for later use.

-- Place two of the posts (which are now 7.5cm x 10cm x 457.5cm) on the ground so that they are 305cm apart at the bottom and touching at the typ. Where they touch, plane off a few inches so that the two pieces meet evenly over an area of from 7.5cm to 10cm. Secure temporarily to prevent movement.

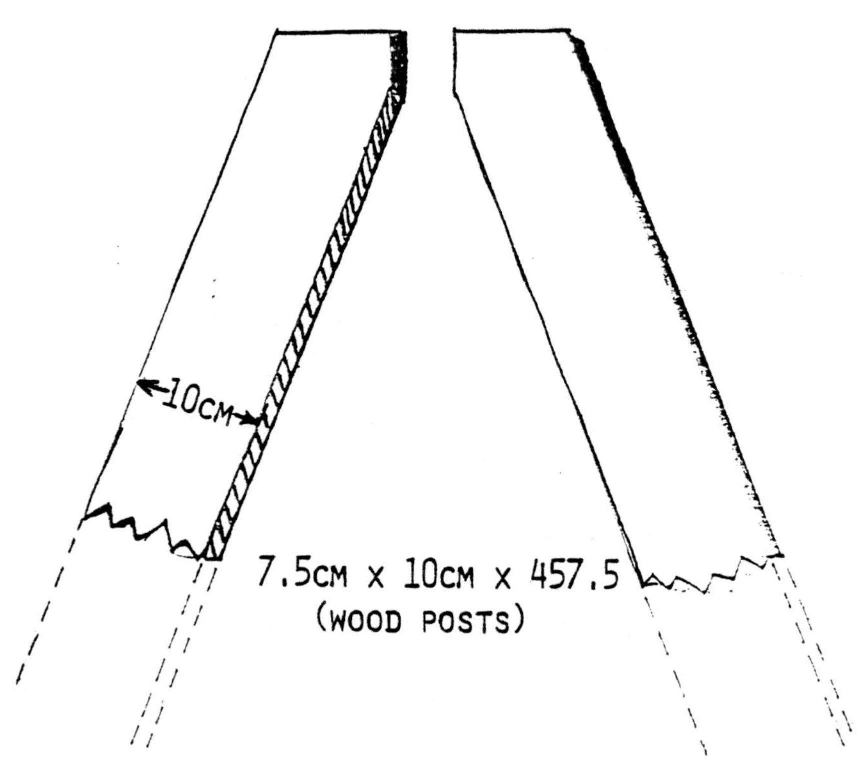

-- Repeat this same procedure with the two other posts.

-- Cut the two pieces of 2.5cm x 10cm x 427cm wood in halves. These pieces will be used as braces.

-- Attach the braces to the A-frame sections as shown. Bolt permanently the two braces attached to the A-frame section which will be opposite the position of the connecting rod. Nail braces on the other A-frame section until it is clear that connecting rod will not strike it when in operation. Check to <u>make sure</u> that the legs of the frame have remained 305cm apart while the nailing process was going on.

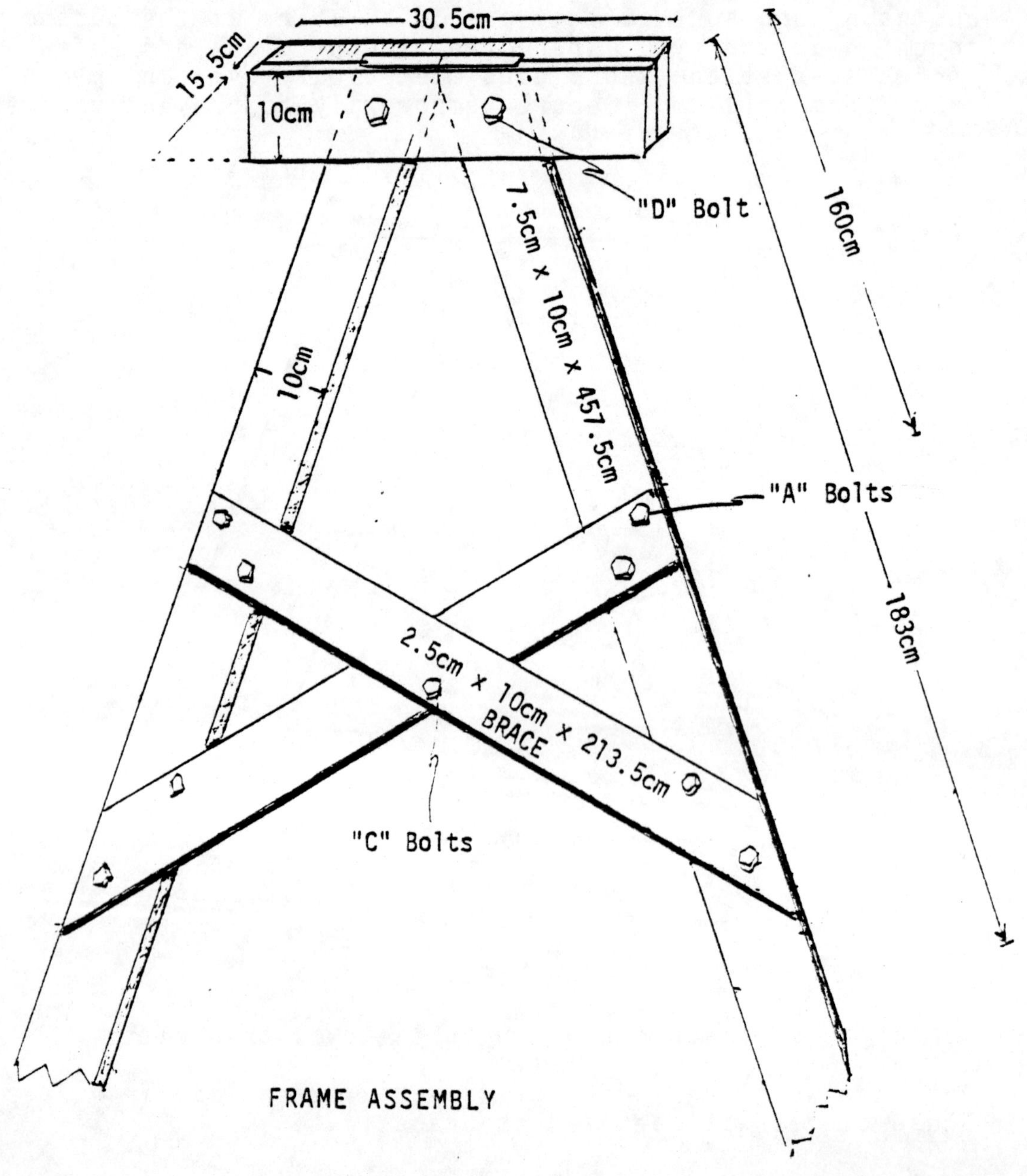

FRAME ASSEMBLY

-- Fasten the braces to the frame between 160cm and 183cm from the top of the frame.

-- Take the 30.5cm blocks which were cut from the four, 7.5cm x 10cm x 488cm wood posts. Cut recesses in them so they almost touch around the top of the A-frame.

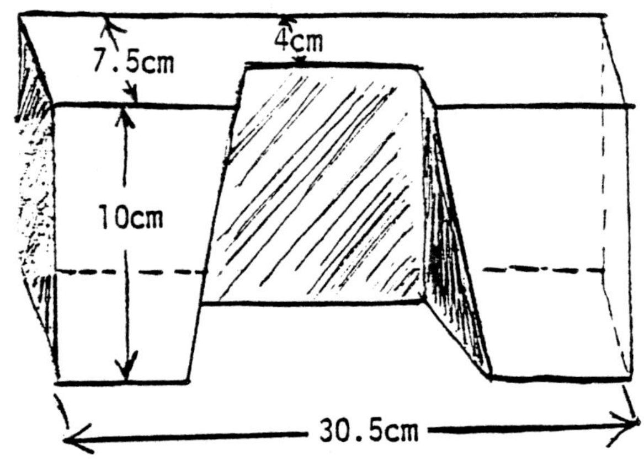

WOOD BEARING BLOCK

-- Drill the holes and secure the wooden bearing blocks with (D) carriage bolts. 2 bolts will suffice. The bearing will be attached to the inside wooden bearing block after the frame is erected.

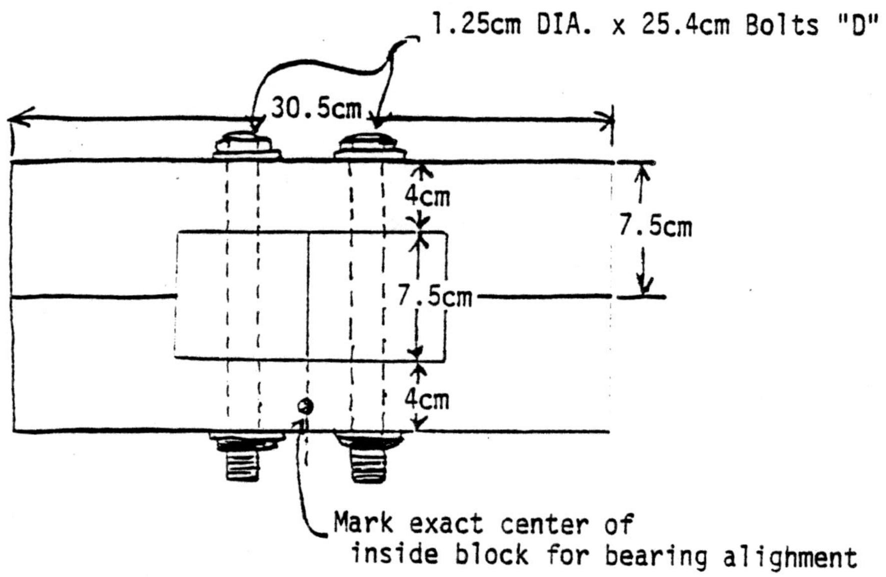

TOP VIEW--BEARING BLOCK ASSEMBLY

-- Mark the <u>middle</u> of the inside wooden bearing block when the blocks are in place. Measure the length and then the width of the block to find the exact middle of both measurements. This mark will be used to position the bearings.

-- Repeat this procedure for the other A-frame.

Erect the Frame

-- Nail all four 2.5cm x 10cm x 366cm pieces to the one A-frame as shown below.

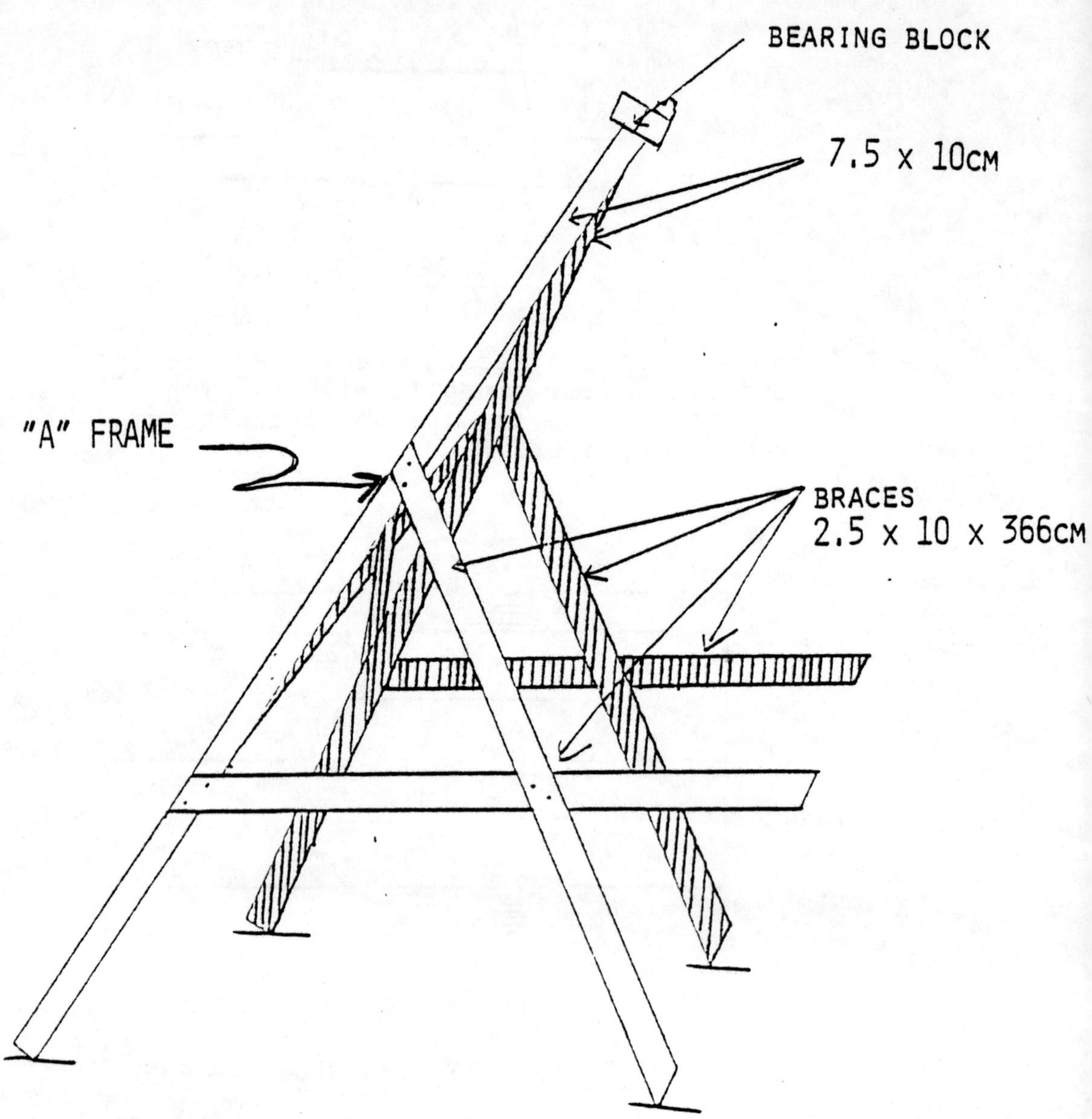

REMEMBER: Do not nail a brace any closer to the top than 160cm and any lower from the top than 183cm

-- Bring the other A-frame up to meet the side braces. Check that the distance between all legs is 305cm and the distance between the tops of the A-frames is 80" (203cm) as measured between the marks made earlier on the exact middle of the inside wooden bearing block.

-- Secure these braces which attach the two A-frames with (B) carriage bolts.

-- Join center crossing with (C) carriage bolts.

-- Add stability by placing 2.5cm x 10cm board around the entire structure. Keep these braces 160cm below top of A-frame.

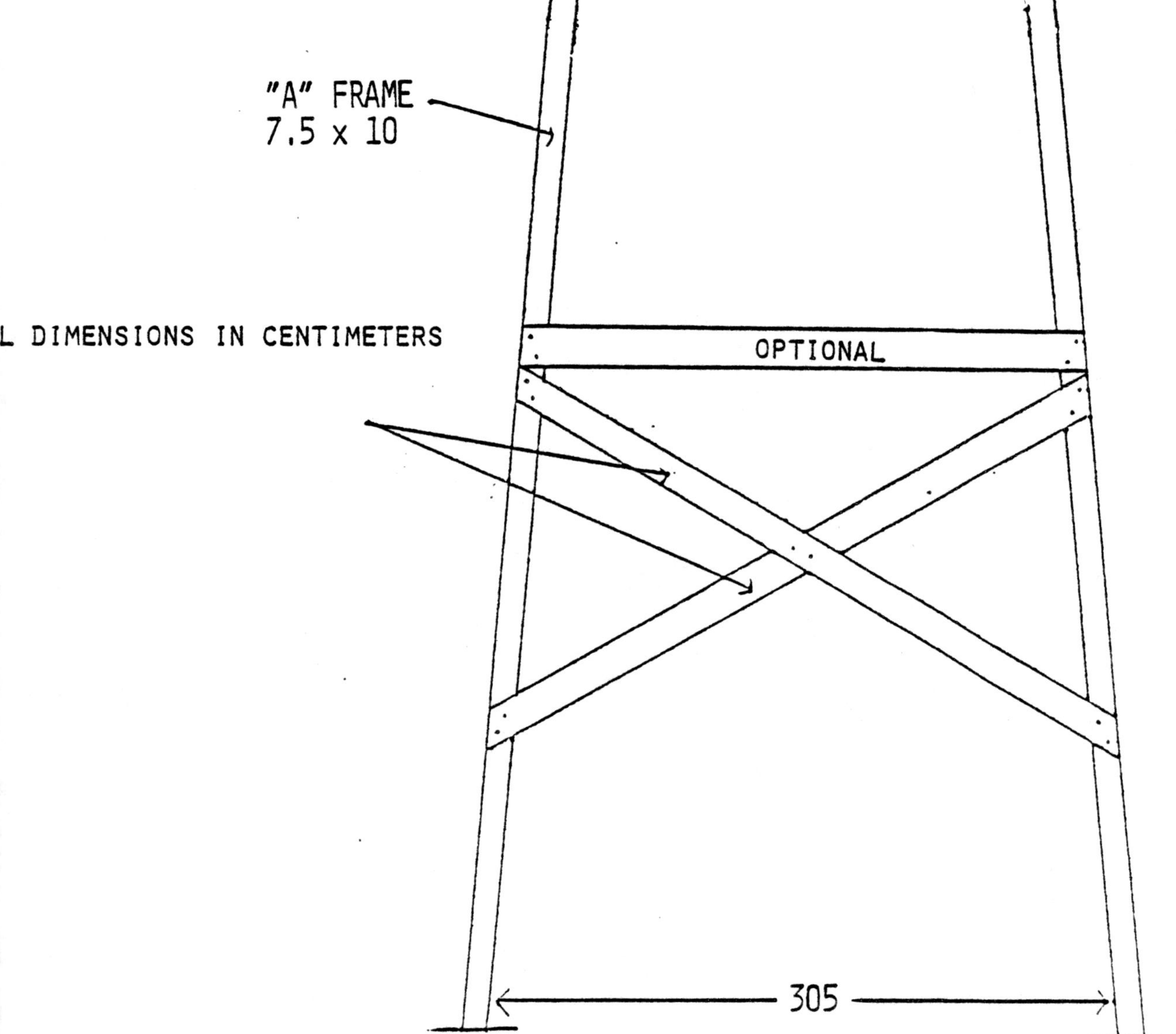

-- Place the erected frame on the foundation. With the self-aligning bearings, chalk, string and level, in hand climb to the top of the A-frame.

-- Find the center of each wooden bearing mount on the top of each of the A-frames.

-- Attach the string so that it passes between the two points and secure it slightly beyond. This is the alignment for the bearing.

-- Use the level to insure that the bearing sits square. Shims can be placed beneath each bearing if needed to level the bearing.

-- Drill the two 1.25cm holes with the bearing centered on the wooden block and aligned with your string.

-- Make sure the bearings are facing inward while drilling.

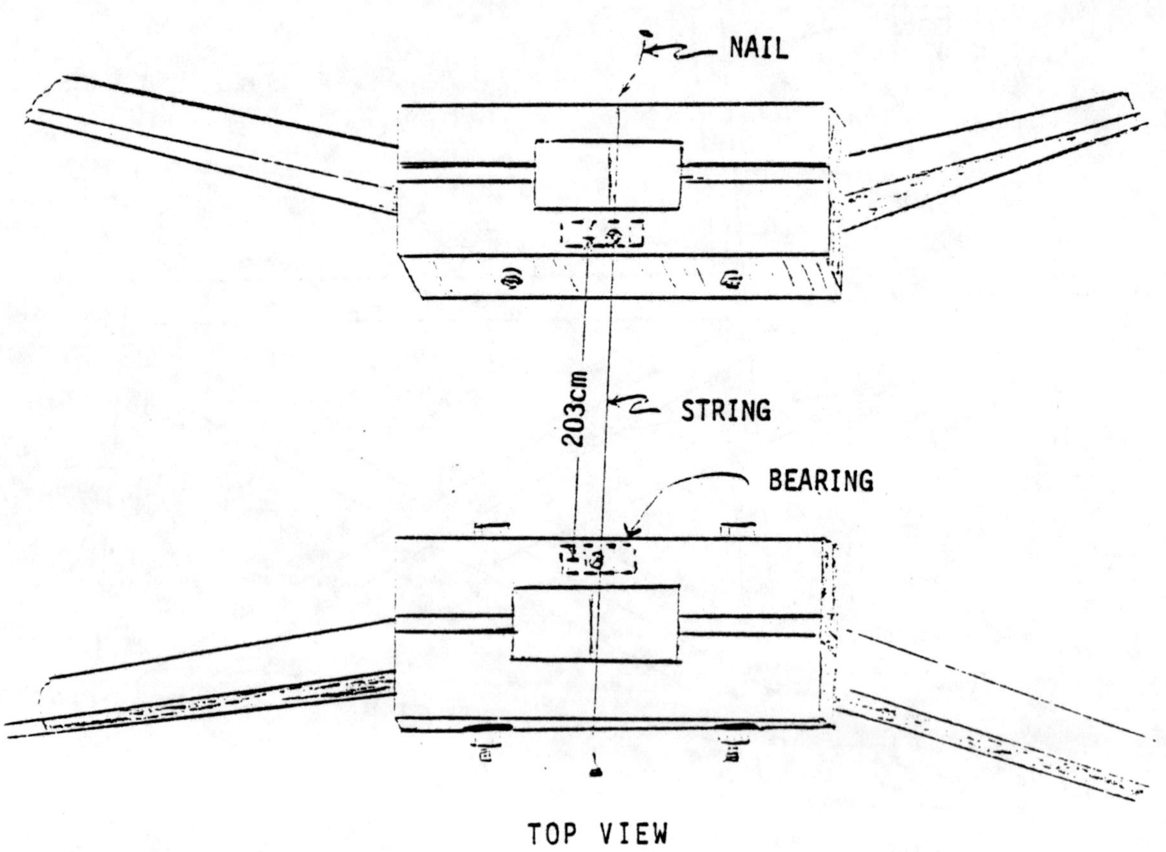

TOP VIEW

Construct The Axle

-- Cut the 1.25cm round bar into 61cm lengths.

-- Cut five pieces of 2.5cm angle bar into 61cm lengths, and the rest into 305cm lengths. There are now 5 pieces of 61cm x 1.25cm round bar, 5 pieces of 61cm x 2.5cm angle bar, and 5 pieces of 305cm x 2.5cm x 2.5cm angle bar.

-- Use the 61cm pieces of angle bar and round bar to construct the axle. Following directions and measurements carefully: this is a critical piece.

-- Use electric arc welding. Make all welds strong.

-- Measure off 35.0cm from both ends on the 2.5cm diameter steel axle rod. Mark and then measure off the remaining center portion into 43cm lengths. This will give welding points (as shown).

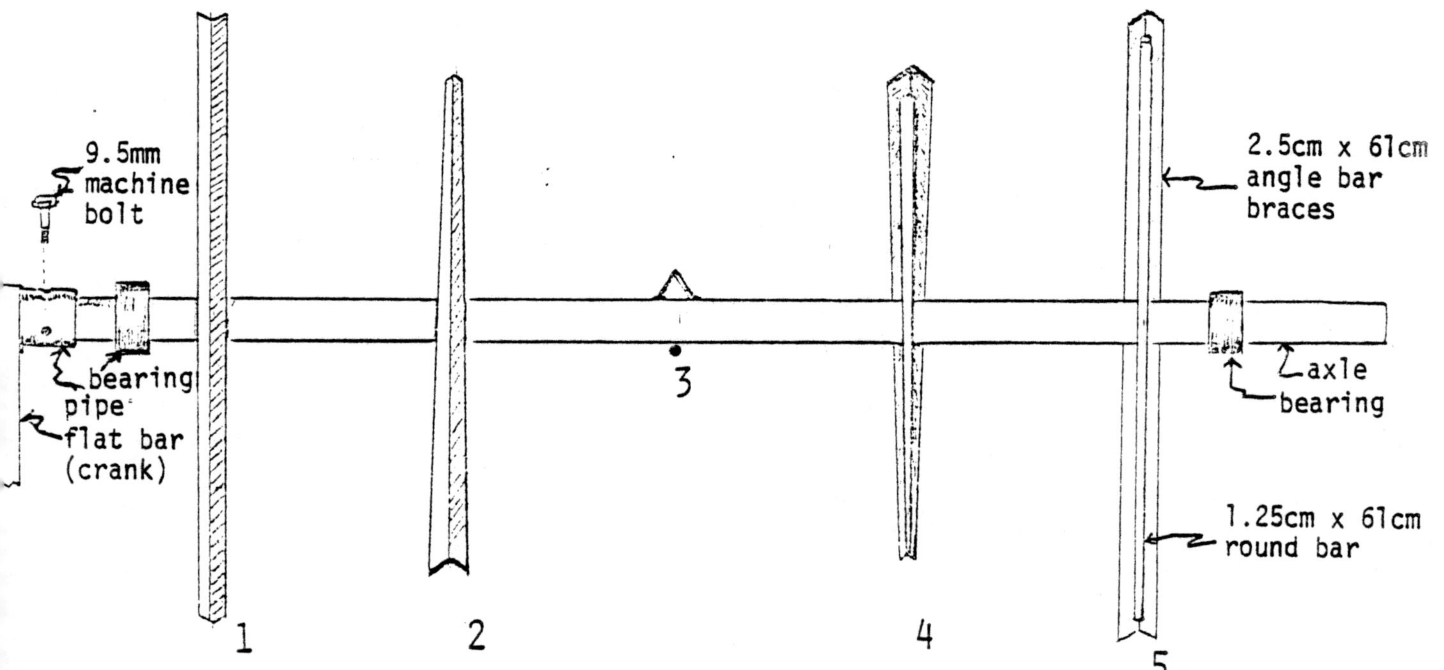

ALL MEASUREMENTS IN CENTIMETERS

-10-10-15- 43 - 43 - 43 - 43 -15- 20 -
5mm

-- Take the first 61cm angle bar length and weld it (centered) on the first spot (1 above).

-- Keep the open side of the V toward the axle.

-- Repeat at spot five placing the angle bar parallel to number one but on the reverse side of the axle.

-- Weld the #3 angle bar (center spot) perpendicular to #1 and #5. Weld the #2 angle bar 45° from the first bar (but still at right angles to the axle!), arranged so the angle bars seem to "walk" around the axle from spot #1 to spot #3. Weld the #4 bar 45° off of #5, but 90° off from #2, not parallel to it, so the bar continue their "walk."

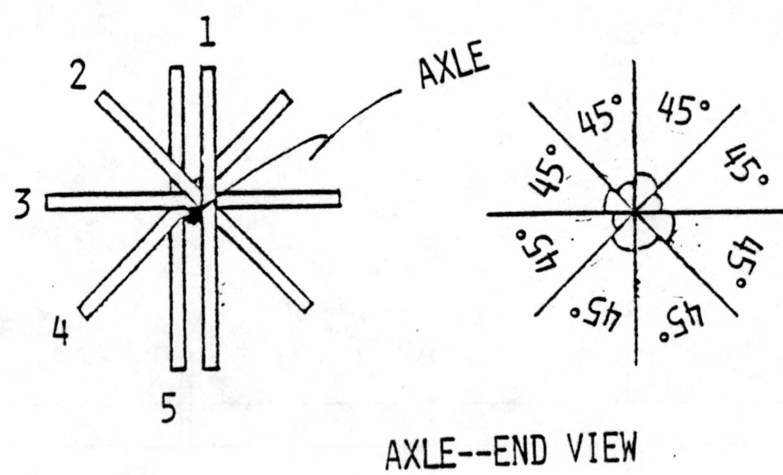

AXLE--END VIEW

-- Bend the 1.25cm diameter bars first to fit, then weld the center and two ends.

-- Weld all round bars in place as shown. (If you plan to construct a number of windmills, it might be easier to construct a wooden rig to get this spacing. Plans for this rig are given on page 27.)

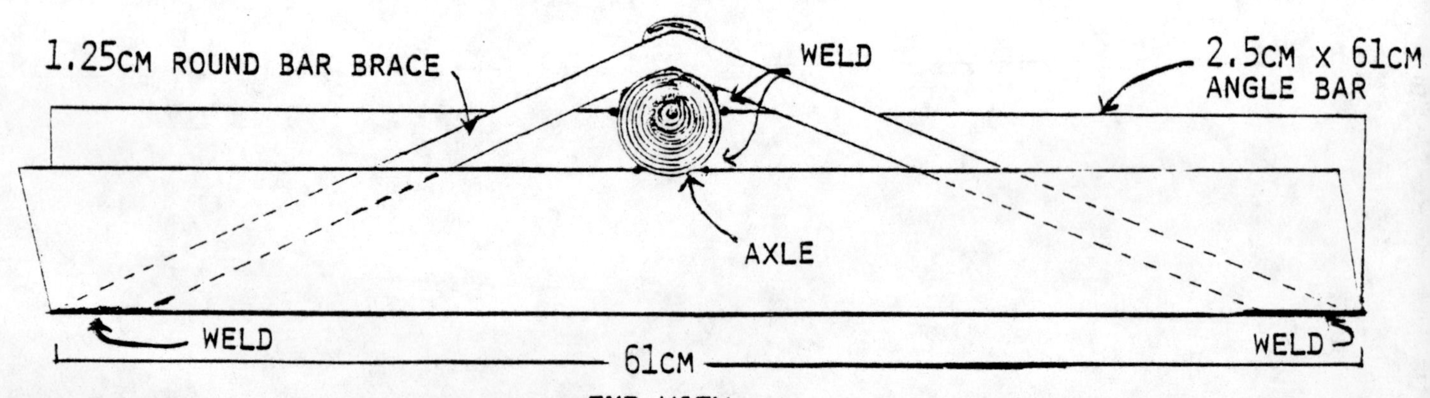

END VIEW

Construct The Spokes

-- Weld 1" x 3/8" (2.5cm x 9.5mm) diameter cap screws to the remaining 305cm angle bars. Place the cap screws as shown.

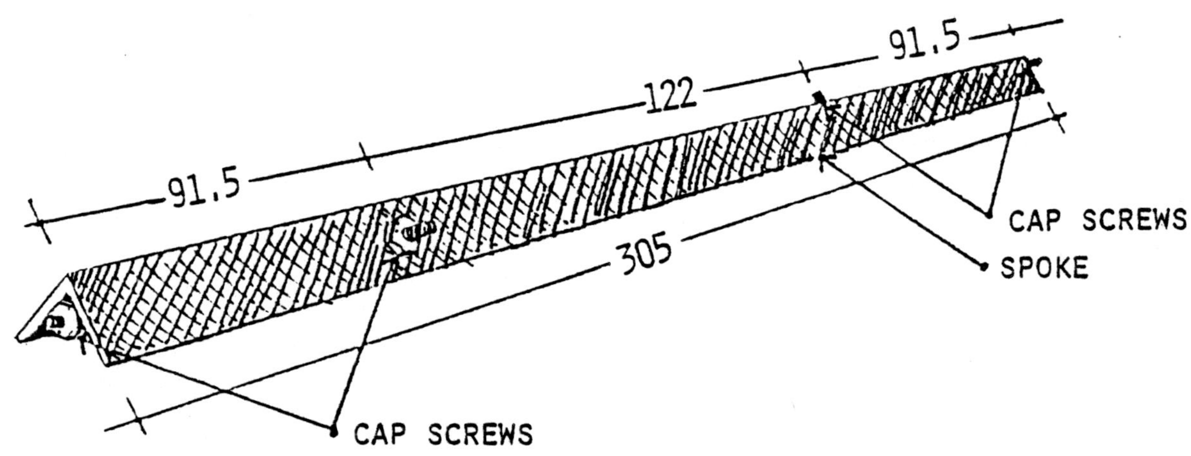

-- Make sure the cap screws are welded tightly. There should be four cap screws on each 305cm angle bar.

-- Attach the spokes (the long angle irons) to the angle iron already welded to the axle with "U" bolts.

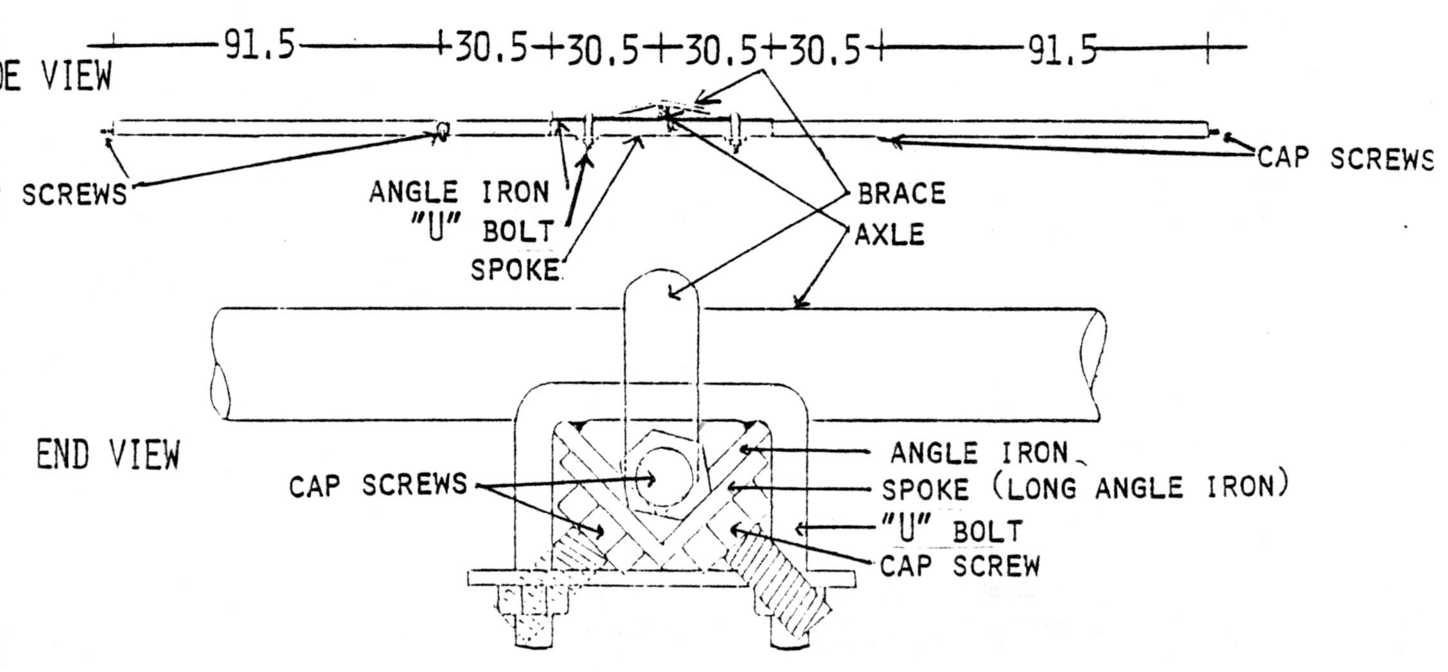

Construct The Ribs

-- Take the 6.5mm round bar and cut it to the proper lengths for ribs. The outer ribs will span the outside circumference of the spokes. The inner rib will span the inner circumference of the spokes. The sails will be attached to these ribs.

-- Make eight long ribs and eight short ones.

-- Bend each rib at the end so as to make a loop. The loop should be small enough to attach to the cap screw, using a lock washer.

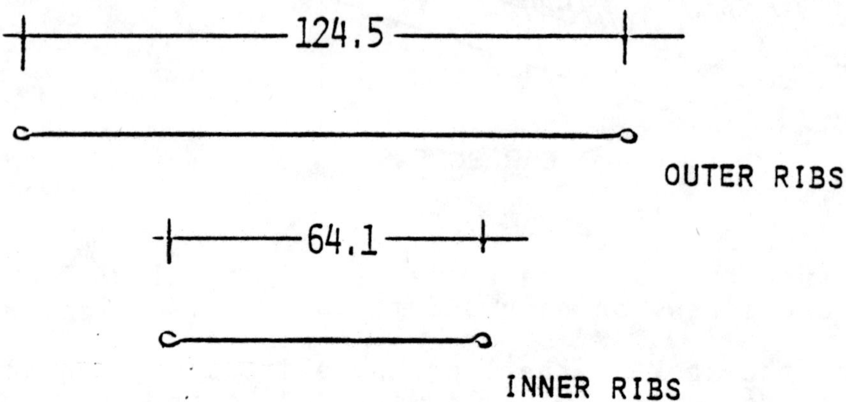

Construct The Canvas Sails

-- Preshrink the canvas and rope before cutting and drying. Do this by soaking it overnight in water and then drying in the sun.

SAIL CUTTING PATTERN

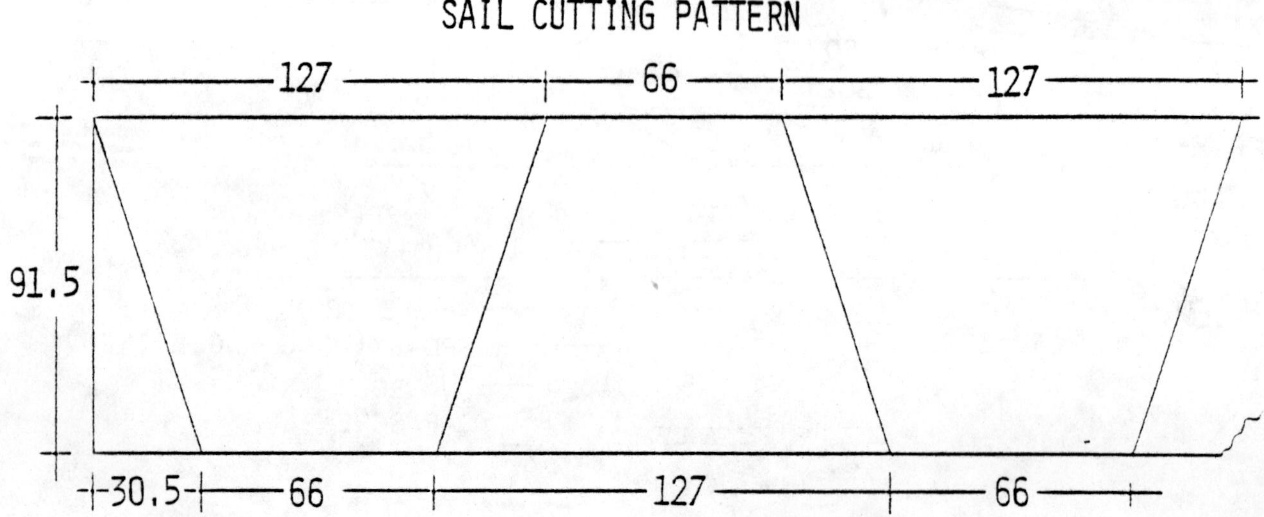

-- Use excess canvas to reinforce the seams.

-- Sew the rope into the seams to make them stronger, and place the grommets (eyelets).

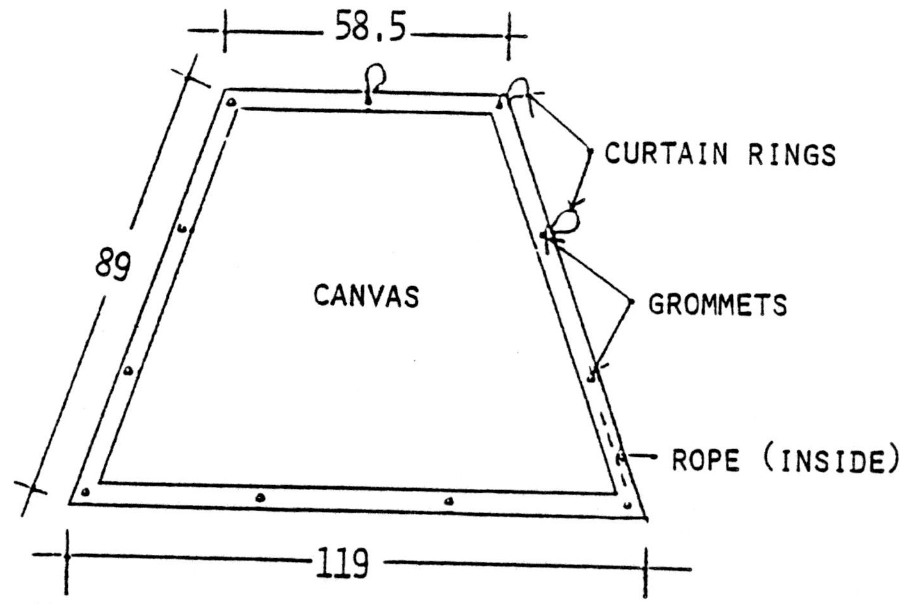

-- Cover both sides of canvas with canvas protector or paint for longer life of the sails.

-- Attach the sails to the spokes and ribs, (See Photo) by clipping the curtain rings around them.

GUIDE WIRES

-- Place guide wire (16 gauge) across the face of the canvas to support it and act as a wind brace.

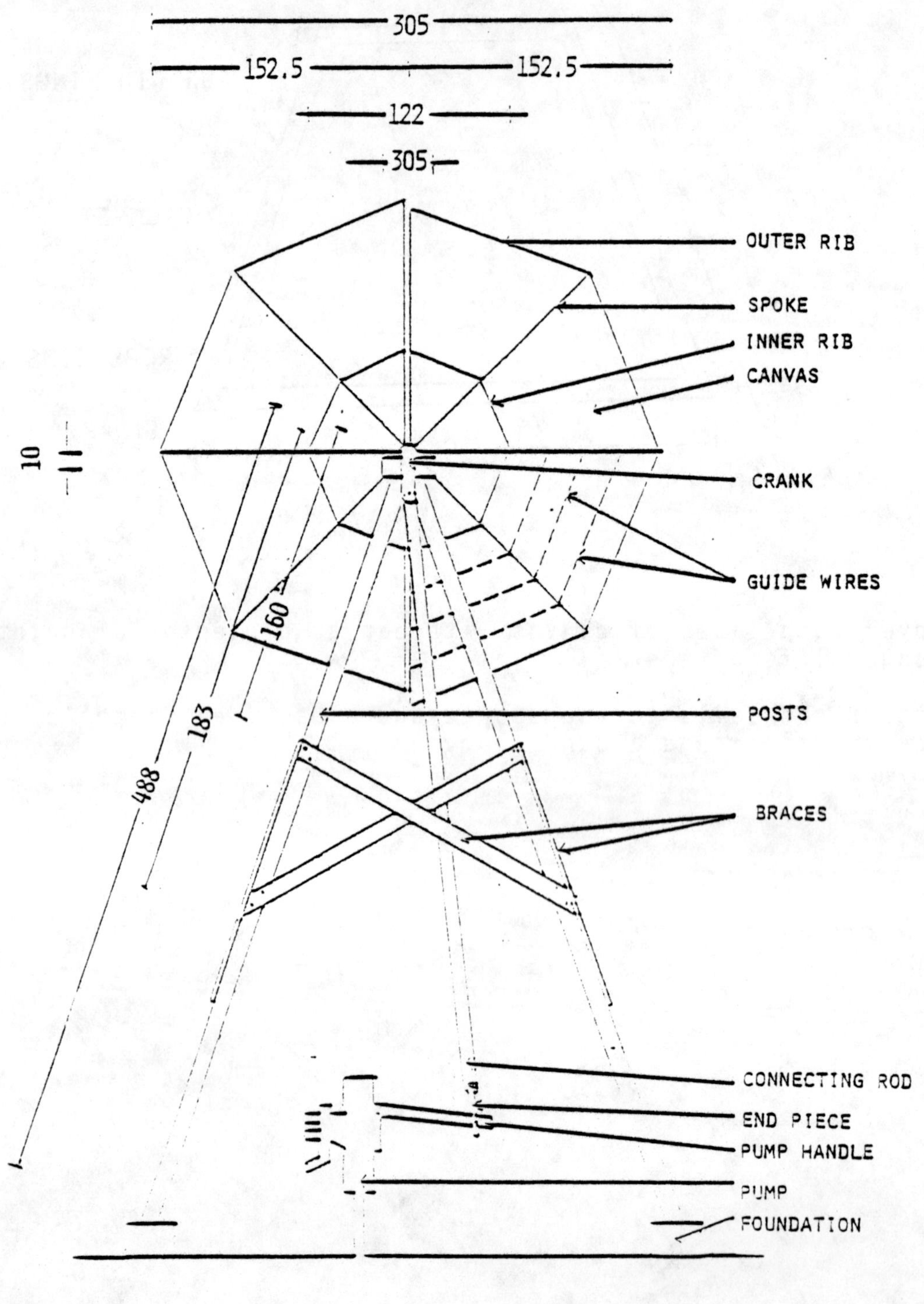

FRONT VIEW

Tension Wires

-- Place tension wires (18 gauge) at the ends of #1 and #5 spokes (both ends). Run the wire to #3 spoke, and to the axle. (See photo) This is the last step before completion of the windmill axle unit.

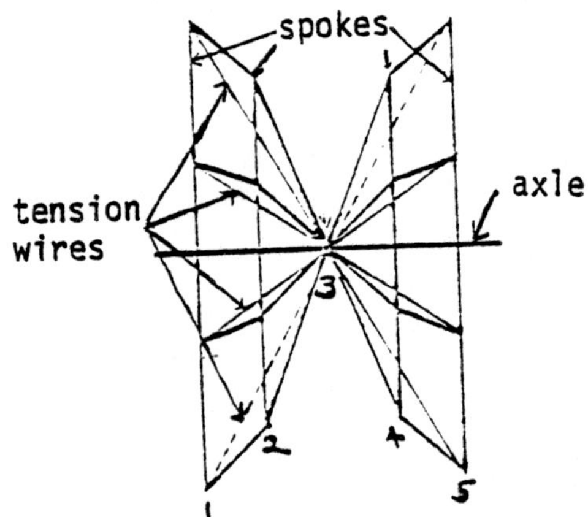

ABOVE: TENSION WIRE ARRANGEMENT

Construct The Connecting Rod Assembly

-- Cut 9.5mm x 5cm x 90.5cm flat bar into three pieces of 20cm each and one piece of 30.5cm.

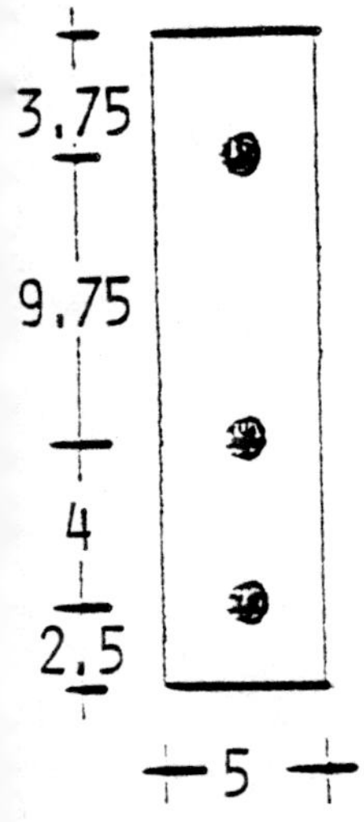

3.75
9.75
4
2.5
5

B" AND "C"
Z BARS LIKE THIS)

-- Drill a 1.6cm hole in the third 8" bar.

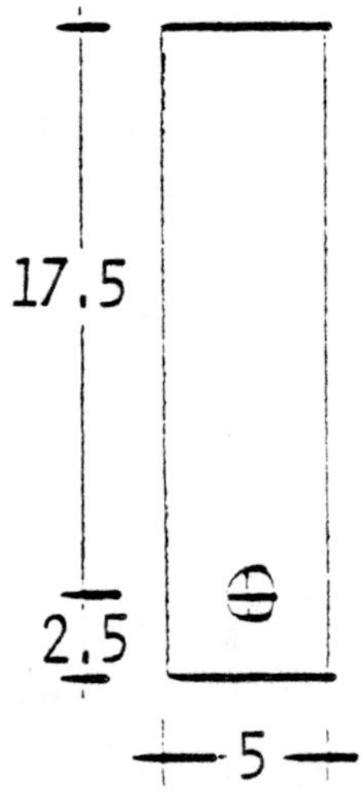

2.5
10
30.5
7.5
7.5
2.5
5

17.5
2.5
5

FRONT

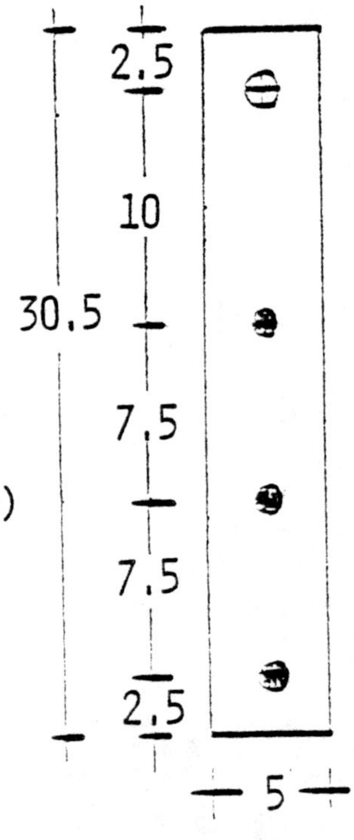

-- Drill 4 holes in the 30.5cm bar.

-- Take the ball bearing assembly and weld it very carefully around the 1.6cm hole in bar 'A'. Make sure the inside diameter will accomodate the 1.6cm machine bolt that goes through it. This bearing assembly may be a replacement part from a Honda motorcycle. However, if such a bearing is not available, substitute by using two additional layers of flat bar with 1.6cm hole drilled through. However, this will act only as a bushing, and may not last as long as a bearing assembly.

-- Weld a 10cm piece of 2.5cm inside diameter pipe onto opposite end of the 1.6cm hole which has been drilled in bar 'D'.

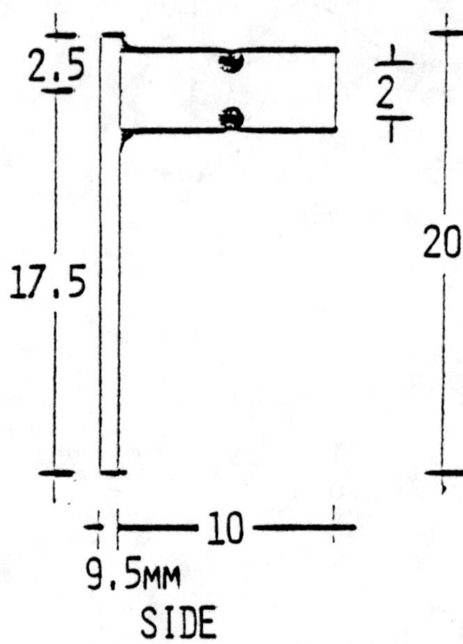

-- Drill in the pipe three equally spaced 9.5mm holes to hold the locking bolts. The locking bolts (E) can be any replacement part available, though they should not be too long. Another way to secure the axle to the pipe is to drill a 6.5mm hole directly through the axle and place a nail or cotter pin through. However, this is not as strong and will not last.

-- Use the 5cm x 5cm x 488cm wood for the rods. This piece of lumber must be at least grade 2 quality. Since it is impossible to determine the exact location from the hole to the pump without first mounting the windmill and pump together, do not cut any excess off until you are sure of the exact distance. This distance is measured from the bottom stroke of the axle to the "down" position of the pump handle.

-- Begin assembling the connecting rod by bolting the 'D' bar to 'A' bar using the 1.6cm machine bolt. The lock nuts lock the bolt to the flat bar 'D' while the bearing assures smooth operation and therefore should not be tight. Keep this critical point well greased, as well as the bolt through the pump handle, and check periodically for wear.

-- Proceed to bolt bar 'A' to wood rod using three of the 9.5mm x 7.5cm machine bolts.

-- Drill two or three 9.5cm holes in the pump handle itself. Using holes closer to the pump will give more water per stroke but fewer strokes; holes further away will give you less water per stroke but more strokes.

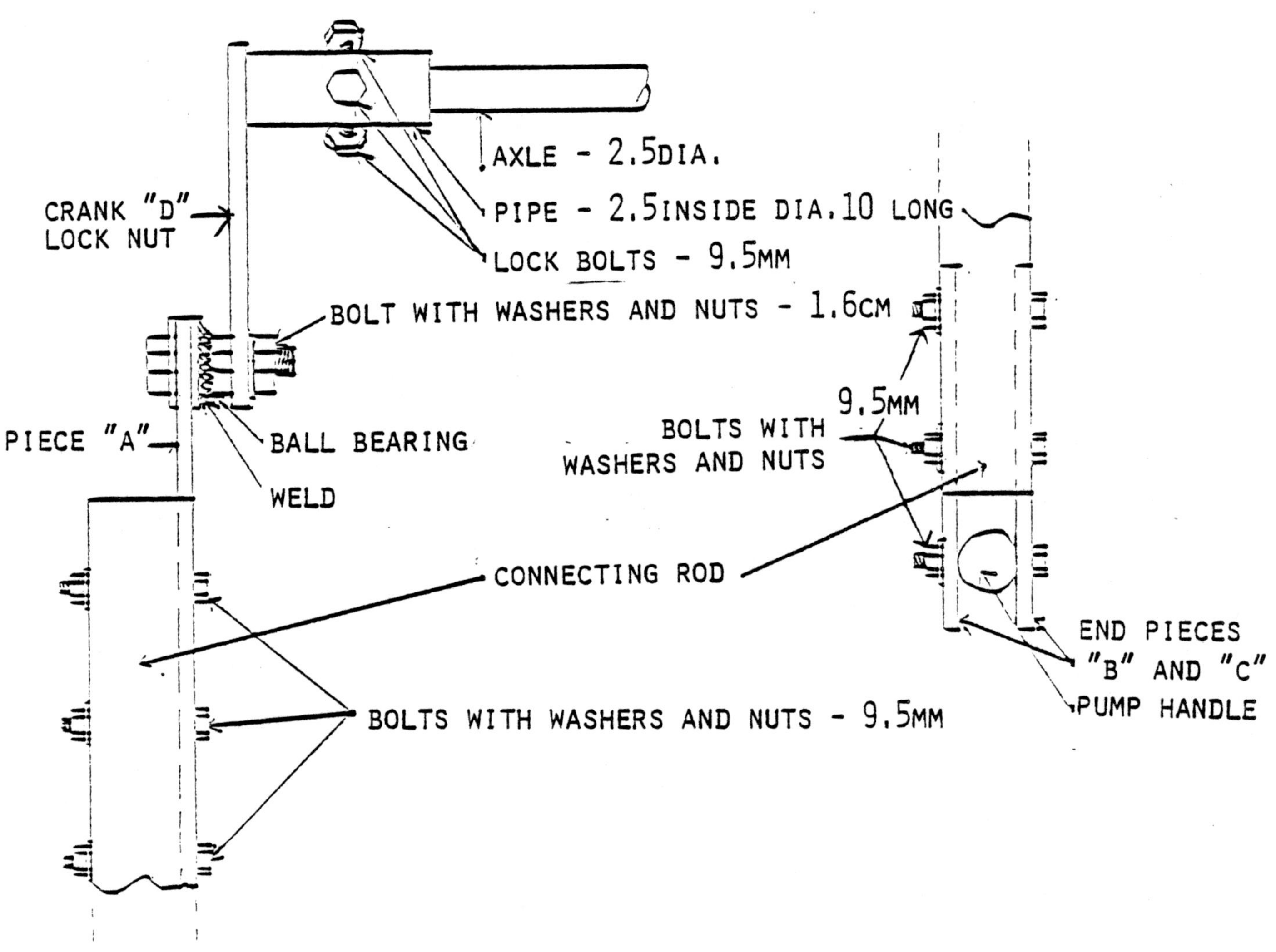

CONNECTING ROD ASSEMBLY

-- Use the remaining three 9.5mm x 7.5cm bolts to attach flat bars 'B' and 'C' to the bottom of the assembly rod and attach the pump handle to the flat bars. This must be done after the windmill axle unit is installed on the frame and the distance is exact.

Assemble Windmill

- -- Place the windmill axle unit on windmill frame. Make sure to bolt the bearings in the holes which were already drilled. Use two ____ x ____ x 12cm carriage bolts for each wooden bearing block. Make sure the washers and lock nuts face inward.

- -- Nail a board near the top of the windmill frame from one A-frame to the other to raise the windmill. Throw a rope over and attach to axle unit. Hoist up carefully.

- -- Attach connecting rod assembly to axle.

- -- Bolt the braces with the remaining 'A' carriage bolts. There should be no obstruction between the braces and the assembly rod.

- -- Operate in a 4-6mph (6-9 kph) breeze. All metal parts should be covered with lead paint and all wood with coal tar.

The windmill is very efficient. Be certain to have an overflow pipe returning extra water to well if a reservoir is used--or the windmill may pump the well.

Footings

Once the windmill is assembled the footings or foundation can be poured to anchor the windmill to the site. Be sure to align the frame to the water pump.

-- Hammer a wooden stake into the soil to mark position at each leg of the frame.

-- Move frame enough so that 40cm square holes 60cm deep can be dug.

-- Dig four holes 40cm square, 60cm deep.

-- Place 10cm wide boards 60cm long over the holes; move and center frame on the boards over the holes.

-- Make sure that the frame is level at all four corners. You may have to place rocks under the boards to make it level.

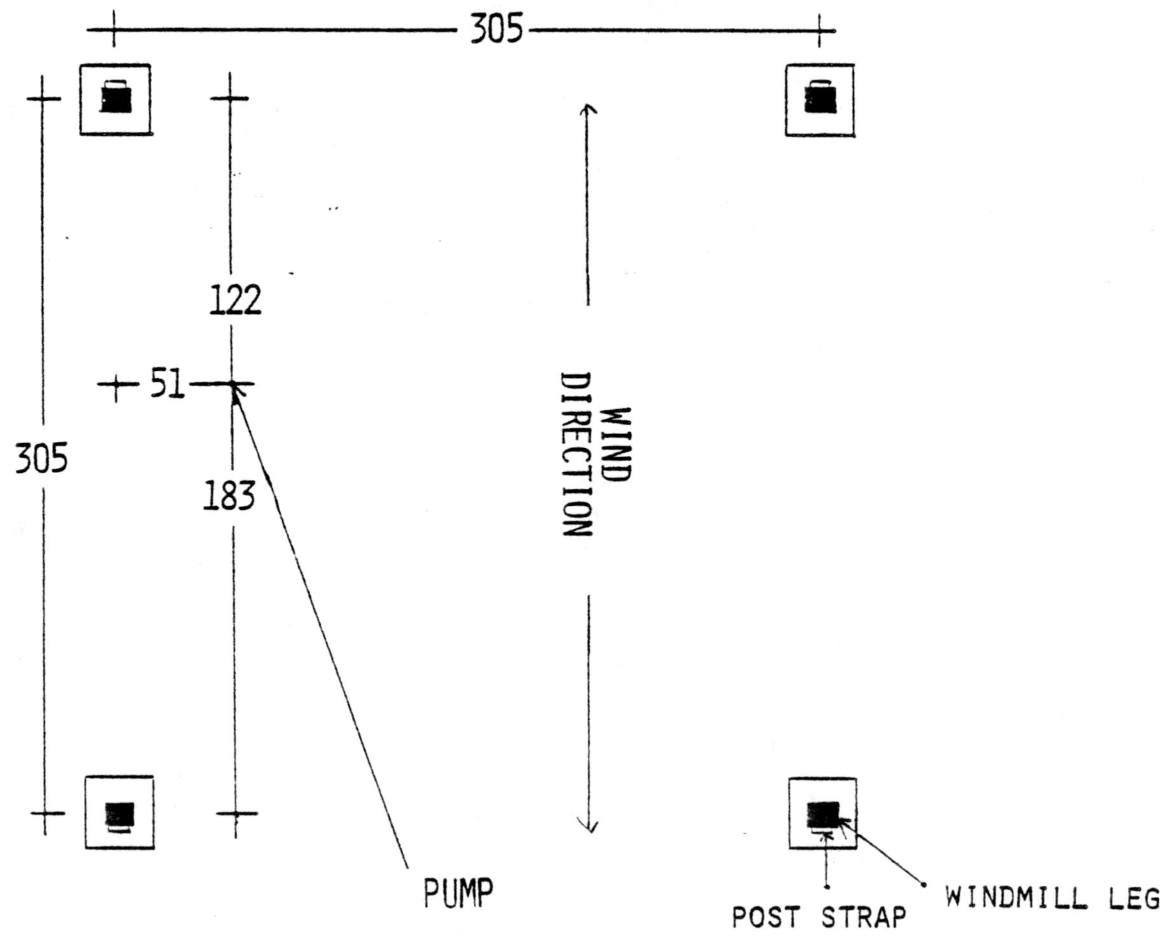

FOUNDATION PLAN

-- Mix concrete in a ratio of one part cement to 2.5 parts sand to 5 parts gravel. Add enough water to mix ingredients into a thick mud-like consistency.

-- Pour concrete into the holes up to the bottom of each 7.5cm x 60cm x 5cm board.

-- Insert 6.5mm x 5cm x 60cm steel plate into the wet concrete 30cm deep. Move the steel plate so that it matches the angle of each leg.

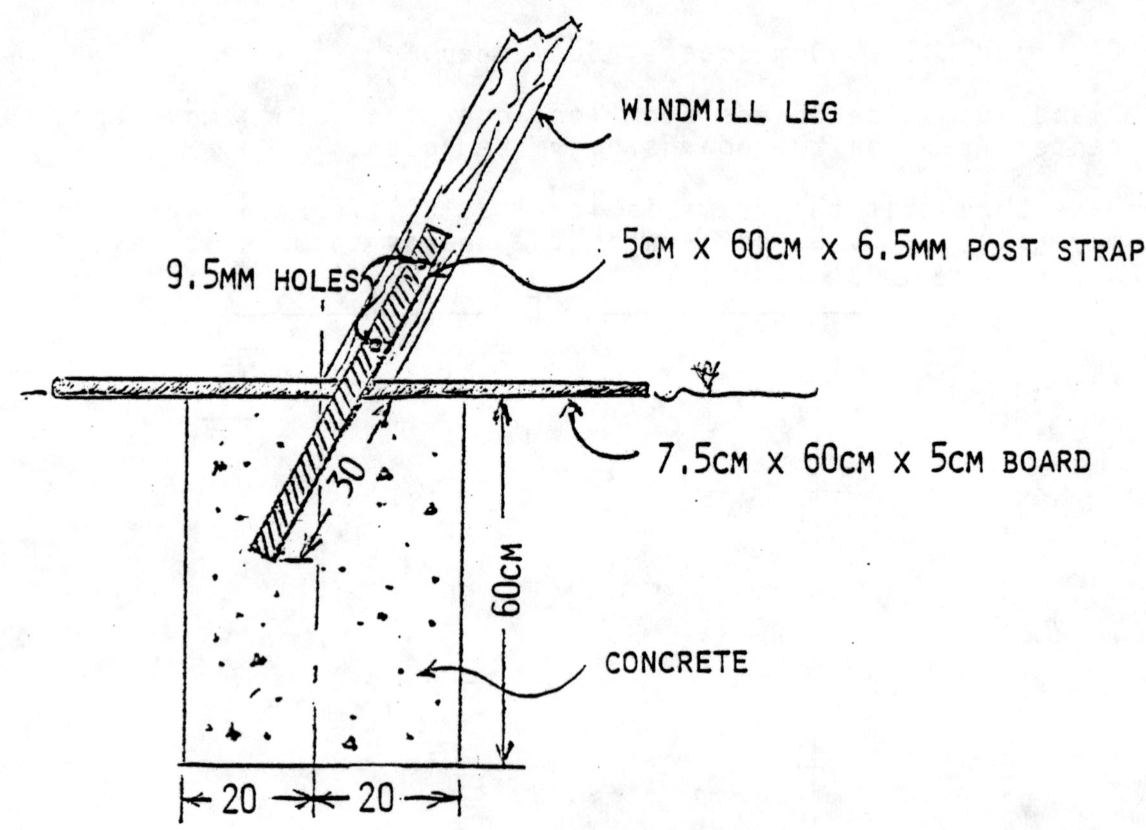

-- Allow the cement to dry for 2 to 3 days. Tilt frame and remove 7.5cm x 60cm x 5cm boards with a sledge hammer.

-- Re-position frame and drill two 9.5mm holes through the metal plate and leg of the windmill.

-- Attach each leg to metal plate using 9.5mm diameter 12cm long carriage bolts.

Welding Rig

If you plan to construct more than one windmill, you may want to build a welding rig like the one shown in Figure 11. The use of the rig can cut down considerably on the construction time.

PARTS LIST

<u>Wood</u>

-- 5cm x 10cm x 13m
-- 2.5cm x 10cm x 9.5m
-- 1.5cm plywood 122cm x 152cm (approx.)
-- Special "C" clamp (includes 1.9cm angle bar 10cm long)

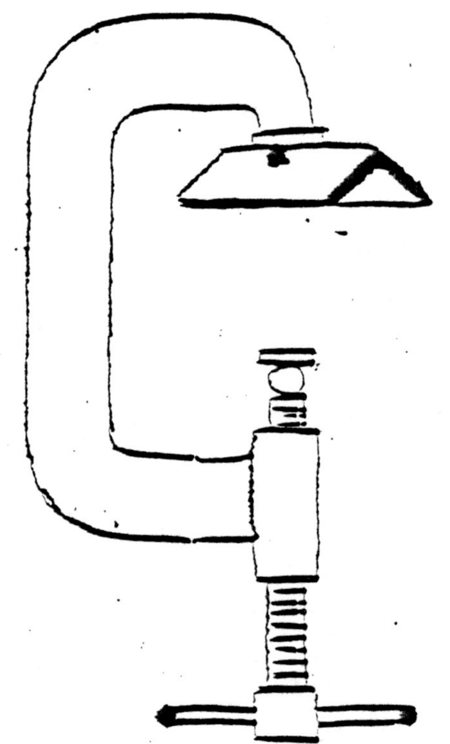

SPECIAL "C" CLAMP -- STANDARD (15cm) C-CLAMP WITH ANGLE IRON WELDED AS SHOWN TO FIT OVER SHAFT.

This fixture does need exacting construction as all stations must be parallel with each other except for station #3 which must be 45° from the rest. If available, scrap lumber may be used, or if desired, angle bars instead of wood. Wood expands and contracts somewhat with changes in temperature and humidity, but should be all right for the purpose intended.

PROCEDURE TO USE THE FIXTURE

1) Place a sail support arm in guide slot in station #2 and push it against the stop.

2) Place the main support shaft in the shaft guide slots in stations #1 and #3 so that the end of the shaft is exactly 35.5cm from the center of station #2.

3) Lock arm and shaft together with special "C" clamp.

4) Tack weld as shown in the welding schedule. Remove the clamp after the third tack and complete #4 as shown.

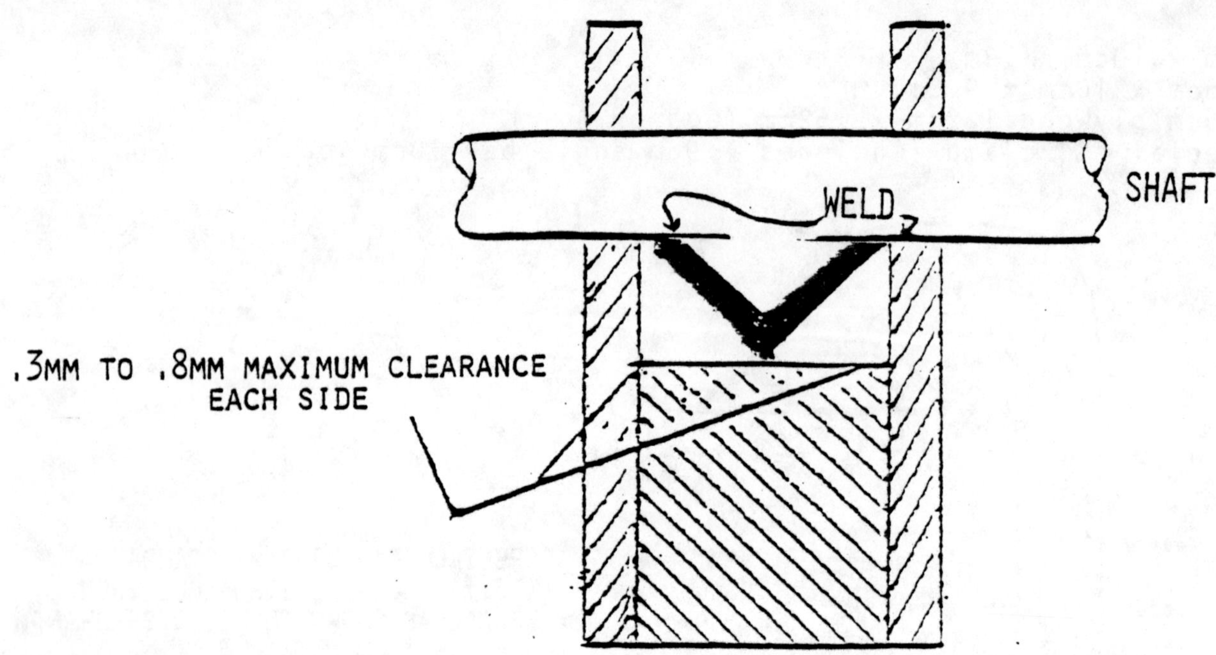

END VIEW OF STATION 2 WITH ANGLE AND SHAFT IN PLACE

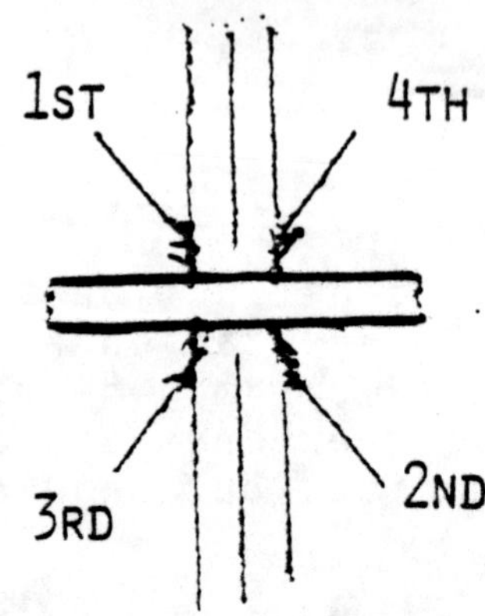

TACK WELDING SCHEDULE

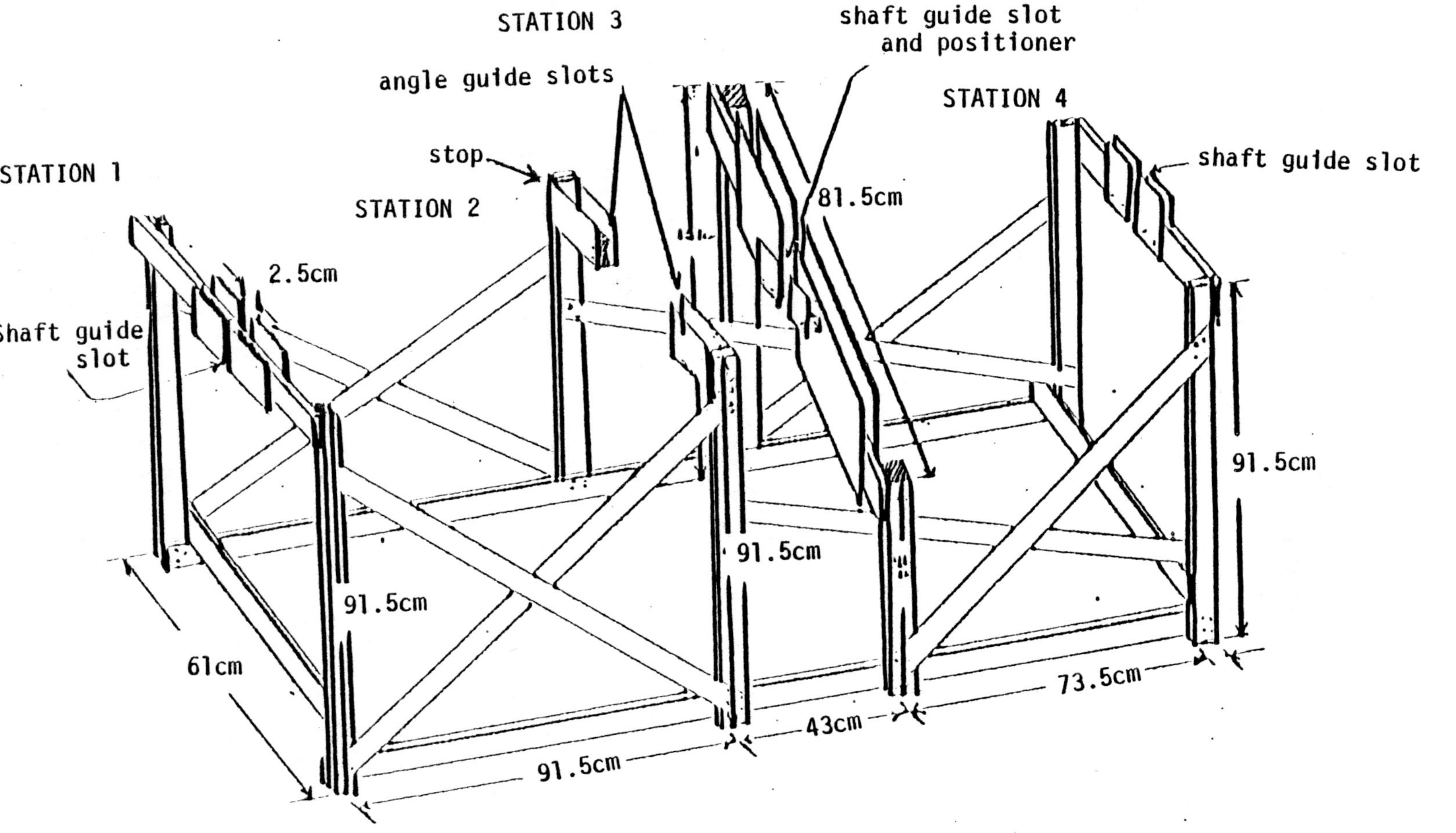

WELDING FIXTURE

All Horizontal and Vertical sections--5cm x 10cm
All Braces--2.5cm x 10cm
Member at Station 3 is at 45 to Horizontal (5cm x 10cm)
Boards forming guide slots can be 1.5cm thick plywood.

5) Repeat the weld schedule as in step #4 above, but apply enough weld metal to make a finished welded joint.

6) Remove the shaft and arm from station #2, rotate 45°, and place in station #3. Slide another arm under the shaft in station #2 up to the stop and clamp the arm and shaft together.

7) Repeat steps 4, 5, and 6 until all arms have been welded.

USEFUL DATA & DESIGN FORMULAS

POWER IN THE WIND

Power in the wind = Velocity^3 X Area

Velocity = wind speed

Area = area swept by the windmill blades when turning (πr^2, where r = radius of wind rotor)

If we use actual units of power, wind speed, and area, we get:

$$\text{Power} = .0012 \times V^3 \times A$$

where Power is measured in foot-pounds per second (ft-lbs/sec)

V (wind speed) is measured in feet per second (ft/sec)

A (area) is measured in square feet (ft^2)

If we want to measure power in horsepower, instead of ft-lbs/sec, we need to know that:

1 horsepower = 550 ftlbs/sec

Therefore,

$$\text{Horsepower} = \frac{.0012 \times V^3 \times A}{550}$$

$$\text{Horsepower} = .0000022 \times V^3 \times A$$

With any known wind speed and rotor diameter, we can now figure the amount of horsepower in the wind passing through the machine. Only a part of this power can be captured by a windmill. The theoretical limit of the power that can be extracted by a windmill is 59% of the power in the wind. Considering the friction losses and other inefficiencies in the windmill, we can expect the windmill to capture only 10-30% of the energy in the wind.

POWER OBTAINED BY THE WINDMILL

If we take the formula for horsepower and include an efficiency factor E (a measurement of the efficiency of our windmill in coverting wind power into pumping power), we can get a realistic figure for the actual power we will obtain at a particular wind speed:

$$\text{Horsepower} = .0000022 \times (KV)^3 \times A \times E$$

where V = wind speed

K = a constant to adjust the units of V (if V is in mph, K = 1.47; if V is in ft/sec, K = 1.0)

A = swept area of the windmill, measured in ft^2 (=πr^2 where r = radius of windmill rotor)

E = efficiency factor (for practical purposes, our design has an efficiency factor of about 25)

Thus the horsepower of the 16.5 ft diameter windmill, in a 20mph wind =

$$.0000022 \times (1.47 \times 20)^3 \times (\pi \times 8.25^2) \times .25 = 3.0 HP$$

NOTE: For metric units the formula for horsepower is:

Horsepower = $.000236 \times (KV)^3 \times A \times E$

where V is measured in meters/sec

$K = 3.28$

A is measured in m^2

MATCHING POWER TO WATER YIELD

How can we measure the amount of power from the windmill required to pump water? Below is a standard table for calculating the water yield that matches the horsepower being produced by a windmill (Table 2). If we know what the wind speed is (see final section), the formulas already given allow us to calculate the horsepower required to pump a given amount of water per hour to a given height. All of this information combined will allow us to go back and calculate the size of the windmill required to pump the water under normal wind conditions.

In an area where a 400-foot well yielding 400 gallons of water per hour would meet the livestock and domestic needs for a rural village, by using either of the charts of Table 2, we see that 400 gallons per hour at a 400 foot depth requires about 1.9 horsepower.

The horsepower a windmill produces at a given wind speed depends on the diameter of the wind rotor. You cannot change the wind pattern in your area; you can only calculate how much power you can expect to get using different wind rotor diameters.

You may find that you need very little horsepower in a low-lift application. You can use the formulas we have presented here to determine approximately how big your windmill should be or how many windmills of the size presented are needed.

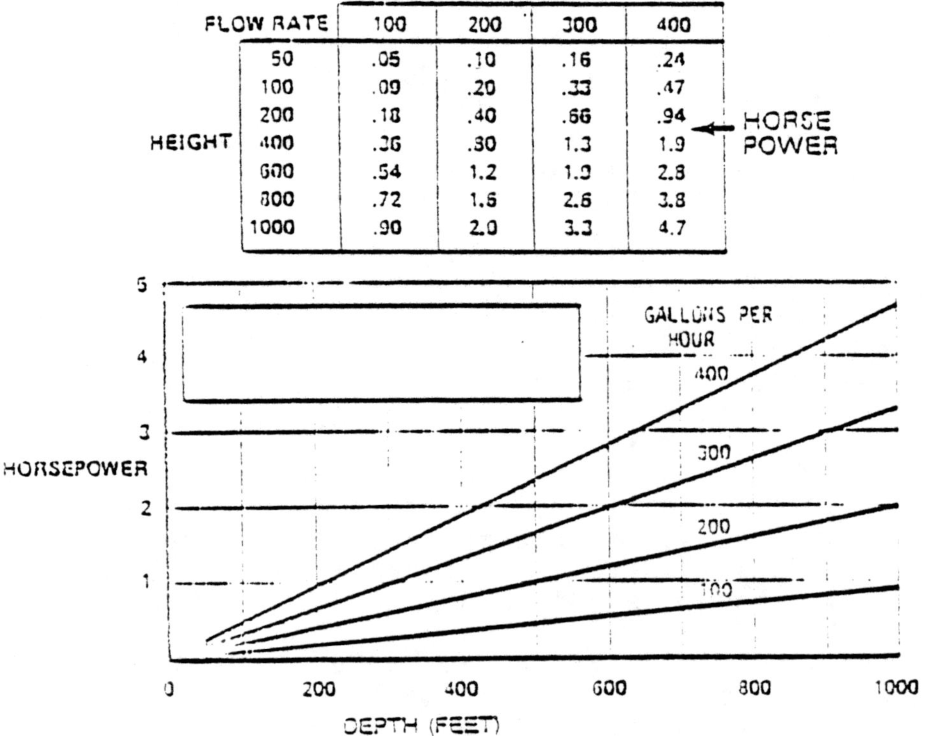

	FLOW RATE	100	200	300	400
HEIGHT	50	.05	.10	.16	.24
	100	.09	.20	.33	.47
	200	.18	.40	.66	.94
	400	.36	.80	1.3	1.9
	600	.54	1.2	1.9	2.8
	800	.72	1.6	2.6	3.8
	1000	.90	2.0	3.3	4.7

← HORSE POWER

Table 2. Horsepower, Depth of Well, Flow Rate

Wind Measurement

The above calculations can be done reliably if you know what the wind speed is apt to be in your area. However, there is very little reliable data presently available concerning wind speed, except at airports, in any country.

An answer to this lack of data has been to use the windmill itself as ameasure of windspeed. The whole unit costs only about $300, and if the wind is found to be insufficient, the upper part of the windmill can be removed and taken to a more promising site. The tripod is then left behind, where it is useful in removing the water pump (operated by some other power source) for repairs.

If you can't afford to buy and maintain an anemometer, you can make one that will give you some idea of how much wind is available (see Figure 33). Some kind of ------------- light-weight (but solid) plastic cups are needed. If these cannot be found, the next best thing to use are cone-shaped cups, which you may have to make yourself. The counter on this anemometer could be a ------------- plastic bicycle odometer (used to measure the distance a bicycle travels).

The rotational speed of the anemometer is proportional to wind speed, which means that the number appearing on the counter in any time period will be proportional to the average wind speed. The device can be put on top of an automobile placed at the height inteneded for the windmill rotor. The count can be recorded every day or every few hours. If there is a regular time of day when the wind always blows, you may want to measure this separately.

The only difficulty is to match the numbers on the counter with a known measure of wind speed. The easiest way would be to run it next to a borrowed commercial anemometer, and simply compare the results. The production of a number of these low-cost anemometers would be a great help in a regional wind power development project. Since you probably can't borrow a commercial anemometer, you can instead attach your home-built anemometer to a pole tied to a car or truck, when there is no wind. Using the car's odometer (mileage indicator) and a watch with a second hand, you can make several test drives at a steady speed. This will give you a good estimate of the proper relationship between the numbers on the counter and the wind speed. You can also do this with the car's speedometer. For example, you drive a steady 15 miles per hour for 5 minutes. The counter reads 1240. Multiply this by 60/5 to get the expected count for one hour at 15 mph (1240 x 60/5 = 14880). Therefore, one hour of 15 mph wind would produce something close to 14880 on the counter. If you are using a car's odometer with 1/10 mile as the smallest unit (when used with a bicycle) it will require about 80 revolutions to add one unit. Thus you might get only 15 units on the counter after driving 5 minutes at 15 mph. In this case, 180 units on the counter would represent 1 hour of 15 mph wind.

GLOSSARY

abundant -- very plentiful; more than sufficient.

access hole -- a hole in one part that allows you to get inside to change or lubricate another part.

adequate -- sufficient; enough.

aligned -- to be in a straight line.

alignment -- arrangement in a straight line.

alterations -- changes.

anchor pins -- pieces of metal which are placed in the concrete foundation and connect the windmill tower legs to the foundation.

anemometer -- a device for measuring wind speed.

axes -- more than one axis.

axis -- a straight line on which an object rotates.

axle -- a rod on which a wheel turns.

back pressure -- pressure in the opposite direction to the flow of water.

barren -- without much plant life.

bearing housing -- a pipe that holds the bearings in position.

bore hole -- a hole in the ground made by drilling, for a pump.

boring -- drilling.

brace -- support.

brazed -- joined by melting metal.

burrs -- tiny rough pieces of metal.

butt weld -- to weld end to end.

car jack -- a device which can lift the heavy weight of a car a short distance off the ground.

clamp -- hold together firmly.

clearance -- open space between two parts.

constraints -- limitations, boundaries.

convert -- change into.

counter balance -- a system in which one weight balances another, making both easier to move.

cross bar -- a bar or pipe which connects two other pipes.

cross braces -- supports which go across the main supports.

cross bracing -- a set of supports that go across and strengthen the main supports.

cross section -- a view of a part as if that part were cut in half.

data -- information.

deflection -- movement away from the normal position.

dia. -- diameter.

diaphragm pump -- a pump that operates as a flexible wall collapses and expands.

dies -- tools used in cutting threads on steel rod to make bolts.

drill press -- a workshop machine for making holes at a precise angle (usually 90 degrees) in metal or wood.

drive mechanism -- the part of a windmill that operates the pump; in this windmill either the eccentric wheel or crank alteration.

duration -- the time that a thing continues.

eccentric wheel -- a wheel in which the axle is not at the center point; in this windmill, as the wheel turns, the axle moves up and down.

enlarge -- make bigger; make larger.

fabrication -- construction; manufacture.

fittings -- small parts used to join, adjust, or adapt other parts.

flange -- a collar on a wheel to hold it in place and give it strength; specifically, a collar on the eccentric wheel of this windmill.

flux (brazing) -- a substance used to help metals fuse together.

foundry -- the process of melting and molding metals.

friction-drive -- a power system in which power is transferred from one part to another by forcing the two parts to rub together.

friction losses -- power (energy) losses due to the rubbing of parts against each other, creating small amounts of heat.

galvanized -- metal covered with zinc to protect it from rust.

grease (verb) -- to lubricate; to put grease on something.

grease nipples -- small metal fittings through which grease can be pumped; for example, to lubricate bearings inside a pipe housing.

grinding wheel -- a workshop tool that has a wheel with a hard, rough side used to smooth or remove small amounts of metal from a tool or part; also, the wheel portion of this tool.

gusset -- a triangular metal brace for reinforcing a corner or angle.

gust -- a sudden, strong rush of air.

gusting -- blowing in sudden, strong rushes of air.

guy wires -- cables that strengthen the sides and keep them in position.

hacksaw -- a hand saw for cutting metal.

housing -- a frame, or pipe for containing and supporting some part or mechanism.

hub extension -- a piece of pipe which sticks out from the front of the hub and provides a place to attach guy wires to strengthen the blades.

inertial forces -- for this windmill, forces which tend to make the windmill remain stopped if it is already stopped, or tend to make it keep moving in a particular direction if it is already moving in that direction.

insert -- to put inside something.

isometric view -- a method of drawing figures in which three dimensions are shown not in perspective, but all dimensions are shortened to give the impression of proper relative size.

jack (car or automobile) -- a device which can lift the heavy weight of a car a short distance off the ground.

linkage -- a part that connects two other parts.

load -- the weight to be lifted by the windmill pump rod.

load bearing surface -- the surface supporting the entire weight of the windmill superstructure and the pump system.

lubrication -- application of grease to make slippery or smooth.

maximize -- to make something operate at the highest possible level.

metal-working lathe -- a workshop machine which can cut circular shapes out of metal.

milling -- the grinding, cutting or processing of metal.

modified -- changed slightly.

nipple -- a piece of metal with a small opening through which water or grease can be forced.

obstacle -- barrier; something that gets in the way.

odometer -- a simple counting device that records the distance a vehicle has traveled.

optimum -- best.

options -- choices, possibilities.

parallel alignment -- two axes lined up exactly parallel to each other.

perpendicular -- at a 90 degree angle.

pipe fittings -- small pieces of threaded pipe used to connect two longer pieces.

pitch adjustment -- adjustment of the angle at which the blades are attached.

pivot -- turn on an axis.

plunger -- the part of a diaphragm pump that moves up and down.

precise -- exact.

pump rod -- the steel rod going from the top of the windmill to the top of the **pump mechanism.**

pump stroke -- the distance the pump piston travels between its highest and lowest points.

punch -- to knock a hole in a piece of metal using a hammer and a sharp tool.

radial guy wires -- guy wires connecting the blade tips to the hub extension to strengthen the blades.

reciprocating pump -- a pump in which the movement of a piston up and down forces the water up.

remote areas -- areas difficult to get to from outside.

rigid -- stiff.

rocker arm -- a support mechanism which turns on a shaft at one end and moves up and down at the other end.

rotary motion -- turning, rotating.

rotary power -- power provided by a turning shaft.

sealed ball bearings -- ball bearings in which lubrication is not required-

seasoned hardwood -- hardwood that has been cut and allowed to dry for a long time.

segment -- piece of something.

shear block -- two pieces of wood which clamp two pieces of pump rod; under normal conditions this connects the pump rod pieces, but under greater stress one of the pieces will pull out of the block.

shim -- a thin piece of metal used for filling space or leveling.

smear -- spread around.

snug -- close fit.

socket -- a piece or part into which something fits.

spacer -- a piece of metal used to fill space so that two pieces will come together at the right point, or clear each other.

speedometer -- a device in a car which measures the speed at which the car is traveling.

spot braze -- to braze in small spots.

spot weld -- to weld in small spots.

stress -- force acting on a part that tends to strain or alter its shape.

strut -- a brace; a support which resists pressure.

superstructure -- a structure built on top of another.

swept area -- the circle made by the blades of a windmill as they turn.

tabs -- small pieces of metal that are bent for some purpose.

tack weld -- to weld temporarily with a small amount of material.

tap -- to cut threads on the inside of a hole in a nut, pipe, or piece of metal.

terrain (rough) -- land that is difficult to move through.

thermal winds -- winds caused by the heating of the ground by the air.

threading equipment -- tools that can cut thread into metal rod to make bolts.

torque -- the force that acts to produce rotation; a twisting force.

triangular -- having three straight sides.

tripod -- a support having three legs.

turbulence -- irregular motion of air.

U-bolt -- a U-shaped bolt with threads and nuts at both ends.

unevenness -- roughness, irregularity.

varnished -- covered with a paint-like substance made of resins dissolved in oil; for protection from damage by the weather.

vertical axis -- an axis that follows a line straight up and down.

wear ring -- one of two pieces of steel rod bent to form a ring and rub against each other as the windmill pivots in the wind.

welding rod -- a special rod of metal that is melted during the welding process when two pieces of metal are joined.

wind rotor -- the turning wheel and blades of the windmill.

ENGLISH/METRIC CONVERSION

Metric

Length

1 mm = .039 inch (.003 foot)

1 cm = .39 inch (.033 foot)

1 meter = 39.4 inches (3.28 feet)

1 kilometer = 0.62 mile

Area

1 square cm = .155 square inch

1 square meter = 10.8 square feet

1 hectare = 2.47 acres

Volume

1 cubic cm = .061 cubic inch

1 cubic meter = 35.3 cubic feet

1 liter = .264 U.S. gallon

Weight

1 gram = .035 ounce (avoirdupois)

1 kilogram = 2.2 pounds

Temperature

$^{o}C. = 5/9 \times (^{o}F. - 32^{o})$

Example: How many $^{o}C. = 77^{o}$ F.?

$5/9 \times (77^{o} F. - 32^{o}) = 25^{o} C.$

English

Length

1 inch = 2.54 cm (25.4 mm)

1 foot = 30.5 cm (.305 meter)

1 mile = 1.60 kilometer

Area

1 square inch = 6.45 square cm

1 square foot = .093 square meter

1 acre = 0.4 hectare

Volume

1 cubic inch = 16.4 cubic cm

1 cubic foot = .028 cubic meter

1 U.S. gallon = 3.78 liters

Weight

1 ounce (avoirdupois) = 28.3 grams

1 pound = 0.45 kilogram

Temperature

$^{o}F. = (9/5 \times {^{o}C.}) + 32^{o}$

Metric

English

Energy

100 calories = .396 BTU

100 watts = .134 horsepower

1 kilowatt-hour = 3413 BTU

Energy

1 BTU (British Thermal Unit) = 252 calories

1 horsepower = 746 watts (.746 kw)

Common Abbreviations

millimeter (mm)

centimeter (cm)

meter (m)

kilometer (km)

gram (gm)

kilogram (kg)

Common Abbreviations

inch (")

foot (ft. or ')

ounce (oz.)

pound (lb.)

horsepower (hp)

British Thermal Unit (BTU)

Speed

1 mile per hour (mph) = 0.45 m/sec.

Speed

1 m/sec. = 2.24 miles per hour (mph)

CONVERSION TABLES

Units of Length

1 Mile	= 1760 Yards	= 5280 Feet
1 Kilometer	= 1000 Meters	= 0.6214 Mile
1 Mile	= 1.607 Kilometers	
1 Foot	= 0.3048 Meter	
1 Meter	= 3.2808 Feet	= 39.37 Inches
1 Inch	= 2.54 Centimeters	
1 Centimeter	= 0.3937	

Units of Area

1 Square Mile	= 640 Acres	= 2.5899 Square Kilometer
1 Square Kilometer	= 1,000,000 Sq. Meters	= 0.3861 Square Mile
1 Acre	= 43,560 Square Feet	
1 Square Foot	= 144 Square Inches	= 0.0929 Square Meter
1 Square Inch	= 6.452 Square Centimeters	
1 Square Meter	= 10.764 Square Feet	
1 Square Centimeter	= 0.155 Square Inch	

Units of Volume

1.0 Cubic Foot	= 1728 Cubic Inches	= 7.48 U.S. Gallons
1.0 British Imperial Gallon	= 1.2 U.S. Gallons	
1.0 Cubic Meter	= 35.314 Cubic Feet	= 264.2 U.S. Gallons
1.0 Liter	= 1000 Cubic Centimeters	= 0.2642 U.S. Gallons

Units of Weight

1.0 Metric Ton	= 1000 Kilograms	= 2204.6 Pounds
1.0 Kilogram	= 1000 Grams	= 2.2046 Pounds
1.0 Short Ton	= 2000 Pounds	

CONVERSION TABLES

Units of Pressure

1.0 Pound per square inch	= 144 Pound per square foot
1.0 Pound per square inch	= 27.7 Inches of Water*
1.0 Pound per square inch	= 2.31 Feet of Water*
1.0 Pound per square inch	= 2.042 Inches of Mercury*
1.0 Atmosphere	= 14.7 Pounds per square inch (PSI)
1.0 Atmosphere	= 33.95 Feet of Water*
1.0 Foot of Water = 0.433 PSI	= 62.355 Pounds per square foot
1.0 Kilogram per square centimeter	= 14.223 Pounds per square inch
1.0 Pound per square inch	= 0.0703 kilogram per square centimeter

* at 62 degrees Fahrenheit (16.6 degrees Celsius)

Units of Power

1.0 Horsepower (English)	= 746 Watt = 0.746 Kilowatt (KW)
1.0 Horsepower (English)	= 550 Foot pounds per second
1.0 Horsepower (English)	= 33,000 Foot pounds per minute
1.0 Kilowatt (KW) = 1000 Watt	= 1.34 Horsepower (HP) English
1.0 Horsepower (English)	= 1.0139 Metric Horsepower (cheval-vapeur)
1.0 Metric Horsepower	= 75 Meter X Kilogram/Second
1.0 Metric Horsepower	= 0.736 Kilowatt = 736 Watt

WINDMILL

A SURVEY OF THE POSSIBLE USE OF WINDPOWER IN THAILAND AND THE PHILLIPINES, book, 75 pp. plus appendix. W. Heronemus, 1974 on request from Agency for International Development, Washington, D.C. 20523 USA; or $6.00 from NTIS (Accession No. PB-245 609/3WE, N'74; foreign orders add $2.50 handling charge).

THE HOMEMADE WINDMILLS OF NEBRASKA, book, 78 pp. E. Barbour, 1898 (reprinted 1976) $3.00 from Farrallones Institute, 15290 Coleman Valley Road, Occidental, California 95465 USA.

SAHORES WINDMILL PUMP, booklet, 80 pp. J. Sahores, 1975, $6.00 from Commission on the Churches' Participation in Development, World Council of Churches, 150 Route de Ferney, 1211 Geneva 20, Switzerland.

A WATER-PUMPING WINDMILL THAT WORKS, article with plans, 7 pp. New Alchemy Institute, P. O. Box 432, Woods Hole, Massachusetts 02543 USA, in Journal No. 2, 1974, $6.00 from WETS.

WIND POWER POSTER, 1 large sheet, both sides, $3.00 from Windworks, Box 329, Route 3, Mukwonago, Wisconsin 53149 USA.

HOW TO CONSTRUCT A CHEAP WIND MACHINE FOR PUMPING WATER, leaflet, 13 pp., Brace Research Institute, MacDonald College of McGill University, Ste. Anne de Bellevue, Quebec, Canada, 1956 (revised 1973) $1.25 from BRACE.

PERFORMANCE TEST OF SAVONIUS ROTOR, technical report with charts and graphs of the test results 17pp., M. Simonds and A. Bodek for Brace Research Institute, 1964, $2.00 from BRACE.

WIND ENERGY BIBLIOGRAPHY, booklet, 66 pp., Ben Wolff, 1974, $3.00 from Windworks, Box 329, Route 3, Mukwonago, Wisconsin 53149 USA; or WETS.

ENERGY FROM THE WIND, annotated bibliography, 180 pp., B. Burke and R. Meroney, 1975, $7.50 in USA, $8.00 for foreign orders from Publications, Engineering Research Center, Foothills Campus, Colorado State University, Fort Collins, Colorado 80521 USA.

A Listing of Recommended Resource Materials

-- Savonius Rotor Construction, Jozef A. Kozlowski, VITA, 1977, $3.25. A Comprehensive manual on the construction and use of a Savonius rotor wind machine. Has 2 designs, each one with detailed instructions. Also contains a section on hooking up a generator. A highly practical book.

-- "Is There A Place For the Windmill in the Less-Developed Countries?" paper, 21 pp., M. Merriam, 1972, free on request from the East-West Technology and Development Institute, East-West Center, 1777 East-West Road, Honolulu, Hawaii 96822 USA. This is an excellent paper--important reading for anyone seriously considering the use of windmills in developing countries. The author lists the most inportant considerations in choosing to build a windmill, including the most favorable circumstances for a variety of power needs. Four prevailing types of windmills are briefly examined (the Dutch, American multi-vane, propellor, and Savonius Rotor), and the power and other characteristics of each design are briefly compared. There are several tables, drawings, and appendices with supporting information.

-- "Considerations for the Use of Wind Power for Borehole Pumping," leaflet, 15 pp. AT Unit Report, 1976, send nominal sum for postage to Appropriate Technology Unit, Christian Relief and Development Association, P. O. Box 5674, Addis Ababa, Ethiopia. An introduction to the basic considerations for the use of multi-blade windmills for water pumping. Explains the importance of site selection, rotor design, and the other major components along with the criteria that affect these choices. No plans or detailed information given.

-- Food from Windmills, book, 75 pp., Peter Fraenkel, 1975. £ 1.95 or US $ surface mail from ITDG, 9 King Street, Covent Garden, London WC2E 8HN, United Kingdom. Also available from VITA $7.95. This book is to be recommended to anyone doing work on the design of low-cost windmills for irrigation. It thoroughly covers the work of the American Presbyterian Mission on adapting the Cretan sail windmill for use in an isolated area of Ethiopia.

-- Low Cost Windmill for Developing Nations, booklet with dimensional drawings, 40 pp., H. Bossel for VITA, $2.95 plus postage and handling from VITA. "Construction details for a low-cost windmill are presented. The windmill produces one horsepower in a wind of 6.4 m/sec (14.3 mph), or two horsepower in a wind of 8.1 m/sec (18.0 mph). No precision work or machining is required, and the design can be adapted to fit different materials or construction skills. The rotor blades feather automatically in high winds to prevent damage. A full-scale prototype has been built and tested successfully." Performance data is included. The windmill is best used to transmit mechanical energy, but also can be connected to a generator.

-- <u>Proceedings of the United Nations Conference on New Sources of Energy</u>--1961, Volume 7 on wind power, 408 pp., $16.00 from United Nations, Sales Section, Room LX-2300, New York, New York 10017 USA. The seven-volume "Proceedings" were originally printed in 1964, and reprinted in 1974. They are obviously expensive and hard to find; as of this printing they are still available from the above address. Some major libraries have copies. The relevant material in volume 7 (the only volume dealing with wind power) is as follows: a) Studies of wind behavior and investigation of suitable sites for wind-driven plants (15 articles). b) The design of windpower plants (11 articles). c) The testing of windpower plants (4 articles). d) Recent developments and potential improvements in windpower utilization (13 articles), including: "Small Radio, Powered by a Wind-Driven Bicycle Dynamo," by Stam, Tabak, and van Vlaardingen, 5 pp. "Adaptation of Windmill Designs, with Special Regard to the Needs of Less Industrialized Areas," by Stam, 9 pp. "Windmill Types Considered Suitable for Large-Scale Use in India," by Nilakantan Ramakrishnan, and Venkiteshwaran, 7 pp.